Face-to-Face with Doug Schoon
Volume III

Science and facts about nails/nail products for the educationally inclined.

Volume III contains fact-based answers to questions from actual nail professionals on a range of topics from natural nail anatomy/health to artificial nail coatings and salon services

Based on "Face to Face with Doug Schoon", Episodes 50-74

The information in this book has been updated from the video series and contains significant amounts of new or recently discovered information, as well as, many special topics not found in the video series.

Many images in the book are shown also in 3D.

View these images use Red/Cyan Anaglyph 3D glasses

Ordering information can be found in the Appendix

ISBN: 978-0-9979186-3-2
Library of Congress Control Number: 2016916821

An updated, adaption of the Internet video scripts of:

"Face-to-Face with Doug Schoon"

Episodes 50-74

Volume 3

1st Edition

Categories
(Continued from volume 2)

Those with active subscriptions to "Face-to-Face with Doug Schoon" Internet video series can click on the Episode number and questions below and watch Doug Schoon reply (E-readers only). To get your subscription visit www.FacetoFacewithDougSchoon.com

Episode#: Question#

Monomer Liquid/Polymer Powder

52:3 Is there a way to create a test kit that could identify MMA monomer in nail liquids? I want to give salons a way to determine if the product they use contains MMA.

That has been tried and failed. It is not possible to develop a test kit to detect MMA monomer that is consistently correct. These types of simple tests are prone to error and can misidentify MMA monomer as an ingredient in products when it is not actually in the product. The only definitive test for MMA monomer is called a GCMS test. Such a test could cost about $100-200 to run. If the nail product is properly labeled, it will say that it contains MMA monomer and this same information will be on the Safety Data Sheets (SDS), previously known as Material Safety Data Sheet (MSDS).

As discussed in Volume l, it is a myth that MMA is damaging to the health of those wearing or using MMA based nail products. The real problem with MMA is the damage it causes to the nails. This damage is largely due to poor training and this can be a problem no matter what nail product is used. Those who are poorly trained will use any nail product incorrectly and damage the nails. Poorly trained nail professionals are the bigger problem. In my view, why waste time on trying to stop the use of MMA monomer? Instead, nail companies should provide better training for nail technicians and then let the local government regulators enforce the regulations. That would be much better for the industry. Here is a link to what the Nail Manufacturers Council on Safety says about MMA monomer usage. http://www.schoonscientific.com/wp-content/uploads/2016/08/MMA-information_ENG.pdf

57:3 Aren't all acrylic nail powders just a blend of ethyl and methyl methacrylate? Aren't they basically all the same?

No, acrylic nail powder varies greatly and it is a myth that they are all the same. Wonder who started this myth? No surprise, it was started by those who sell low quality nail powders. Of course, they want to fool nail professionals into using inferior and/or incorrect nail powders. They can't make money if nail technicians use the powder that was specifically designed for use with the monomer of their choice. So, they fooled them into believing a "powder is a powder", which is false. Here are the facts - nail powders are polymer blends and their compositions vary widely. For instance, a nail powder could be a blend of 10% poly ethyl methacrylate, which is the polymerized form of ethyl methacrylate and 90% poly methyl methacrylate. Or it could contain the reverse, 90% poly ethyl methacrylate and 10% poly methyl methacrylate...and every combination in-between. When methyl methacrylate or MMA is polymerized to make artificial nail powders, this is considered a safe and appropriate use for this monomer. This is NOT the same as using MMA monomer as the liquid component in a system. I don't recommend using MMA monomer nail products because they are likely to lead to damage of the nail plate.

It is important to understand that nail powders are often customized "co-polymers". A co-polymer is a polymer containing two types of monomers in their structure and not just blended together. Co-polymers are completely different from those created by physically blending two different types of polymers. Co-polymers are one polymer, made from two different monomers. This means that knowledgeable scientists can create hundreds of powders that all have ethyl and methyl methacrylate and each powder will be different in strength, durability and ease of use. There is an even more important difference, as I discussed in Volume l, it is VERY important to understand that various nail powders contain varying amounts of benzoyl peroxide or BPO. The amount of BPO used depends on the formulation of the monomer liquid component of the system. Some monomer liquids require more BPO, while others need less. The BPO concentration is usually between 1-2%, however there is a VERY big difference between a nail powder containing 1% verses a nail powder containing 2% BPO. A nail powder with 2% BPO contains twice as much BPO as a powder with only 1%.

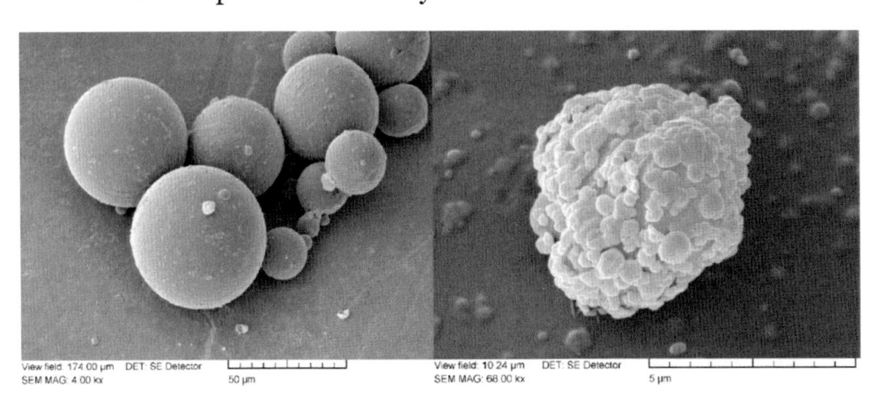

Image 1a/1b: 2D image of acrylic powder particles magnifed 4,000 times (right). Benzoyl peroxide particle magnified 68,000 times (left). Notice that 1b is the same particle seen in the center to the lower powder particle in Image 1a.

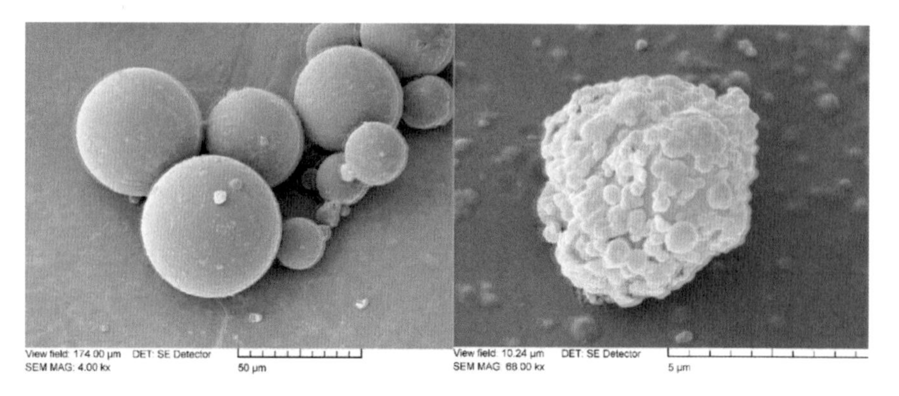

Image 1c/1d: 3D image of 1a/1b (above)
***Red/Cyan** 3D glasses required*

The amount of BPO in the powder determines how "completely" the liquid monomer will cure. Too little BPO, the enhancement will under-cure. Too much BPO, the enhancement will over-cure. Over-curing can lead to discoloration, brittleness, cracking, breaking, chipping and loss of adhesion or lifting. Under cured nail coatings may have a lower resistance to staining or they can be overly flexible and have increased cracking at the stress zones near the free-edge. However, there is a much more important problem to consider, under-cured nail coatings are much more likely to cause adverse skin reactions for clients and nail professionals. Under-curing leaves excessive amounts of monomer trapped in the coating. Therefore, fresh dust and filings may be rich in monomer. Prolonged or repeated contact to monomer-rich dust and filings may lead to skin overexposure, which is a leading cause of allergy to nail enhancements.

If you use the incorrect powder, it may not contain the correct amount of benzoyl peroxide. In other words, don't use a 1% BPO containing nail powder with systems designed for use with a 1.5% BPO powder, and the opposite is true. Even a half percent difference in BPO can lead to an under-cured enhancement and higher risk of skin irritation or allergies. This is especially true during the first hour after creating the enhancement. This is when the coating contains the highest concentration of uncured monomer and the potential for over exposing the skin is at its highest. (See Volume 1, 29:1 for more info.)

57:5 My elderly client has worn acrylic for over 20 years. She has poor circulation and cold hands, so her nails lift a lot. I've tried everything; overlays, new tips and overlays, and warming her hands up before her service and during application. I think she needs to take a "nail break" and recondition her nails, but she doesn't want to. What would you suggest I can do to either improve her wearability or get her to take a break?

I don't know what a "nail break" would accomplish. The nail plate doesn't get "tired" or "fed up" with nail coatings and there is no truth to the myth that nails need to breathe. What is a break supposed to accomplish? Also, I don't understand how poor circulation or cold hands can affect adhesion, so I doubt that this is the problem. I've noticed that nail technicians never suggest giving the nails a break, except as a last resort. When everything else they've tried doesn't work, throw up your hands and tell the client to take a nail break. This seems to be a typical pattern, first blame the products, then blame the client, then tell them to take a break from nails. What happened to this step, "*I should carefully examine my own techniques to see if I'm doing something to cause the problems.*" This is the first thing a wise nail technician will do.

It doesn't matter how long you've been doing nails. In fact, the longer you've been doing nails, the harder it will be to find solutions to difficult problems, unless you challenge your own techniques or methods. Many times, veteran nail techs will fall into a pattern and begin to do "factory nails". Everyone gets basically the same set of nails, rather than to customize the service for each client based on their needs. For example, many (if not most) nail technicians over-file the natural nail. However, older clients nail plates grow more slowly. This results in the same parts of the nail plate being filed repeatedly, before it finally grows over the free edge. In addition, nail plates of older clients often begin to naturally grow thinner. This natural thinning is made worse by filing or excessive buffing. Perhaps what the client in question really needs is less filing and buffing. I've not seen the nails, so I'm speculating, but if this nail tech is doing everything correctly, why would the client need time for her nails to recondition?

File and Buff Less! Respect your client's nails, avoid thinning the nail plate! In fact, that's a general solution to many problems that clients experience, no matter what their age. If nail technicians would work to actively avoid excessive filing and buffing, their clients would have thicker nail plates and would be better able to wear their nail coatings. As the old saying goes, *"You can't build a strong house on a weak foundation"*.

58:3 My nail primer freezes at low temperatures in the bottle and it takes around 30 minutes to completely out thaw. I've heard contradictory information from suppliers. Is a thawed primer ok to use?

This could depend on the formulation of the primer, so if the directions specifically state, "Do not refrigerate", then those directions should be followed. However, in most cases the primer will more than likely be fine to use. Even so, don't open the container until it has warmed completely back to room temperature. Otherwise, water can condense inside the bottle when you open it. When the primer becomes cloudy or forms a solid powder that settles to the bottom of the container, it's time to part with your primer bottle and get a new bottle. I don't recommend refilling and reusing the same bottle of primer, since a waxy substance can build up inside the primer to cause contamination and increase lifting. The waxy material is concentrated skin oils, which may be deposited back to the surface of the plate to adversely affect adhesion.

65:4 Another nail tech told me that if the acrylic liquid smells, that means it must contain MMA, is this true?

That's a reckless generalization. All artificial nail systems based on monomer liquids have some odor. None of them are truly odorless. Even those some claim to be odorless actually have an odor. Some have stronger odors than others, but that means little. You can't determine the safety or usefulness of a nail coating product by smelling it. Many of the best and most useful monomer ingredients used in artificial nail liquids have noticeable odors, so it would be completely incorrect to assume that the odor must be from MMA (methyl methacrylate). It is far better to obtain the Safety Data Sheet or SDS. If the product contains MMA in greater than 1%, it must be listed on the SDS (aka MSDS).

In my opinion, no nail technician should ever use an artificial nail product without FIRST reviewing the SDS. Otherwise, they may miss valuable information related to proper and safe handling. However, when I explained this to this specific nail professional, she replied that she did download some of the SDS from the company website, but this particular one was not available. In my view, that's a red flag and I would NOT use any nail enhancement product unless the SDS was freely available and easy to obtain.

67:4 Is there such a thing as a Universal acrylic nail powder that works with any monomer liquid?"

The short answer is NO! I really love monomer liquid and polymer powder technology (aka L&P). I think it is superior over all other nail systems, due to its tremendous versatility for many types of nails. However, this type of nail coating is more technically challenging to use correctly and there are many things to know so they can be used wisely and safely. Nail Powders (polymer powder) contain varying amounts of benzoyl peroxide (BPO), which controls the curing process. Small changes in the amount of BPO create big changes in how nail products will cure. Too much BPO can cause over-curing which is often seen as overheating, brittleness and discoloration. Too little BPO leads to under-curing which can lead to adverse skin reactions or skin sensitivity. This sensitivity is caused by trapped monomer in the filings left over when too much monomer liquid is in the bead. Prolonged/repeated exposure to this can cause skin sensitivities. Just as UV gels must be properly cured, so must all types of artificial nail coatings, including monomer liquids and polymer powders and colored polymer powders. They need to have the correct concentration of BPO to properly cure. The best way to achieve this is to use the correct powder- the one that was intentionally designed to work with the monomer liquid of your choice. Also, it is very important to use the correct ratio of monomer to polymer- a medium dry bead, never wet or runny. The runnier the bead, the more monomer that's left in the nail coating, as well as the dust created when the nails are filed. Uncured monomer may be trapped inside dust released when the enhancement is filed. Using the wrong powder can alter the amount of BPO in the bead and this can significantly increase the risk of adverse skin reactions for nail professionals, so great care should be taken. In short, monomer liquids and polymer powders are a matched pair that should never be parted- if the goal is to work safely- and that should be the goal for ALL nail professionals.

68:1 My local council is threatening my business because of complaints about odor from residents upstairs above our salon. We have every conceivable ventilation and air purifier on the market. My question is I can't find any data that gives the odor detection level of EMA. Would you know what it is?

If you are using home/office-style air purifiers, those will not work in a nail salon. Something made for bedrooms or offices should not be relied upon in the salon setting. The monomer used to create liquid and powder systems is called ethyl methacrylate or EMA for short. EMA is so easy to smell, that you can smell just one molecule of EMA diluted with one million molecules of air, which in technical terms is called 1 part per million (ppm).

Ventilation is important for all salons. In my view, if you can't afford proper ventilation to ensure the work place breathing air is of good quality, then you should not be performing these services... it is that important. Proper ventilation is a requirement to working safely and all salon and school owners are responsible for ensuring that those in the building are receiving good quality air. There are two companies that I believe provide high quality ventilation systems, "Aerovex Systems" (US and EU), "Filtronics" AB (EU only). Both specialize in salon ventilation, so I recommend contacting them for more information about choosing the right system for your situation.

69:3 In places where it is too cold, salons are heating the monomer in a baby bottle warmer. Is this wise to do?

Warming the monomer up to room temperature because it became cold overnight in the salon can be very different than "heating up the monomer". These products should NOT be heated above normal room temperatures. Heating could cause service breakdown issues such as discoloring. Also, heating would increase inhalation of vapors and could lead to over exposure by inhalation. Avoid warming such products above room temperature. If the goal is to return cold monomer to room temperature, there is a much better way to deal with this issue. That is to prevent the products from becoming too cold in the first place. Some salons get very cold at night after closing, so I recommend keeping products stored inside an empty "ice container" or "chilly bin" overnight. This insulated container will prevent products from getting too cold at night, therefore additional warming is unnecessary. This will help protect the integrity of the products and to avoid related service breakdown issues.

71:3 I've had a client for about one year, soaking off and reapplying a new set of nails about every five weeks without problems, no breaking or lifting. However, lately I'm seeing lifting and breaking in the second week. Do you know what might cause this to happen?

It is hard to know what went wrong, simply from this description, but in my view, going five weeks between services is far too long and just trouble waiting to happen. Rebalances should occur every two-three weeks, depending on nail growth. Four weeks is really pushing it and asking for service breakdown related troubles. Clients that insist on going four weeks often do so to save money, but they can end up with extra problems that could be easily avoided.

UV Curing Products

55:1 Can you give me a simple explanation for how UV products cure?

The simplest way to do this is to describe and explain the four stages of UV curing. This information applies to all artificial nails, to include UV gel manicures and two-part monomer liquids and polymer powders. The first thing you should know is that the curing reactions happen very quickly at first, then slow down dramatically toward the end of the curing cycle. Also, curing and polymerization mean the same thing: each molecule that makes up the reactive ingredients is linking together to form chains which grow longer and longer over time. With that information, let's look at Stage 1 of the curing process.

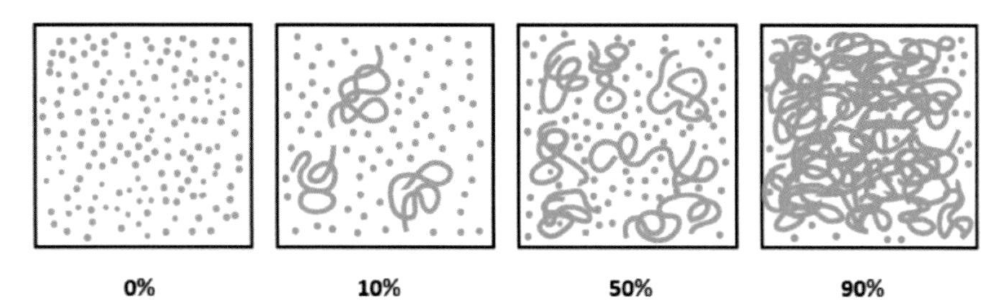

Image 2: Demonstrates the process of polymerization; monomers joining to become part of the lengthening polymer chains. The first image is 100% monomer, and 0% polymer, whereas the final image represents 10% monomer and 90% polymer.

Stage 1 occurs when the product sets or thickens. After 10-15 seconds under a UV nail lamp about 25% of the UV reactive molecules in the formula will chemically link together to create very short chains. In the initial stages, the first short chains that form begin to tangle together and the product becomes very stringy which is due to thickening. This short pre-cure is sometimes used to "lock" in a smile line so that the edge doesn't become blurred when a clear or pink UV gel is added to the nail plate. This is much like when many long hairs tangle to form a knot. At this stage, the UV gel is only about ¼ cured.

Stage 2 occurs when the UV gel becomes at least 50% cured, which is easy to spot. When 50% of the UV reactive molecules link together the UV gel becomes so thick that its surface will harden enough to be shaped with a file. It is VERY important to understand that at this point, the UV gel is only half cured. Hardening of a UV gel does NOT mean it is properly cured. This is one of the most important misconceptions in the nail industry and it even fools many manufacturers who don't understand the science behind the curing process.

Stage 3 is when proper cure is supposed to occur, yet doesn't always happen. During Stage 3, if proper cure is achieved, then the UV gel is cured enough so that the nail coating has the strength and durability needed to perform as intended without service breakdown. You can NOT "see" or "feel" when proper cure occurs, which is why it is so important to follow manufacturer's directions which will help to ensure proper cure. Proper cure is typically reached when about 90% cure, in other words, when about 90% of the UV reactive ingredients are chemically joined together into a strong net-like structure. It is a myth that UV gels cure to 100%. No way! UV gels never 100% cure, nor is this necessary to achieve a proper cure. However, when significantly below 90%, the coating is under-cured. Under curing leads to service breakdown, discoloration or unusual staining of the nail coating, but more importantly this can cause nail professionals and/or clients to develop skin allergies to unreacted ingredients trapped in the nail coatings, as well as their dust and filings. Which is why proper curing is so important to achieve- for the protection of the nail technician and clients.

Stage 4 of the curing occurs "after" the client leaves the salon. By now, the curing process has slowed to a crawl, so only about 5-7% additional curing happens over the next few weeks. Curing slows to a crawl because once the nail coating has hardened, it becomes very difficult for unreacted ingredients to move around and find and join the ends of the growing polymer chains.

UV-A energy drives these chemical reactions for UV curing gels, but in two-part monomer liquids and polymer powders, the heat of the room and the warmth of the client's hand provide all the energy needed to drive polymerization. Therefore, the four stages of curing apply to all types of nail products which cure or polymerize, including adhesives, nail wraps and so-called dip systems. In fact, the only exceptions to this process are traditional nail polishes, since they harden by evaporation of solvents and do not cure by polymerization.

55:2 I am using a 3 in 1 UV gel that is supposed to cure for 60 sec under an LED lamp, but after cure, eight out of ten nails have a little hole on the surface. Does that mean I am over curing or is the product just not of good quality?

Without more information, it is hard to know exactly what the problem is, but my guess would be that the nail coatings were "under cured". Nail coatings become harder and are more difficult to remove when over cured. Whereas under curing creates softer nail coatings that are easier to file and easier to damage. Remember, "3 in 1" means the UV gel is trying to do three different things at the same time. Base coats don't make good top coats and vs versa, so that could be your problem as well. These types of nail coating products are for people who want to save time, not for those who want quality results.

The first thing to do is to make sure you are using the correct nail lamp and if the bulbs can be changed and haven't been changed recently, that may be your problem. Personally, if I were me, I would avoid using a 3 in 1 UV gel. Instead, use a product that doesn't try to be everything all at once. For instance, use a separate base coat and a separate top coat, but make sure they are compatible and designed to work together. Avoid breaking up systems or cherry-picking products from different companies. That's likely to lead to more problems and increase the potential for developing allergic reactions.

55:5 How is the consistency of a UV gel related to the flexibility of the cured UV gel?

The consistency of a UV gel is determined by the consistency of the ingredients used. To explain better, I'll need to discuss how these products cure. Only low power UV energy sources are used to cure UV gels, which means nail lamps emit very low amounts of UV energy, so curing is not very efficient. This helps explain why it is not possible to properly cure the monomer liquids used to create two-part systems- they will not cure in a reasonable amount of time using a nail lamp. To overcome this problem, UV gels aren't based on monomers, but instead they are made from pre-made short chains of monomers called "oligomers".

"Mono" is latin for "one" and "mer" means "units", so monomers are individual "molecules". "Oligo" means "few", so an oligomer is a short chain of monomers, typically up to 3,000 monomers long. Since monomers are small molecules, they move around more quickly and easily inside their container, which is why they are flowable liquids. Oligomers are longer and more bulky, so they get in each other's way and can become temporarily tangled up. Their bulky size slows them down so they can't move nearly as quickly or easily as monomers. The result is that the oligomers have a tendency to be thicker and stringier, which is why they have a gel-like consistency. Depending on which oligomer is used, the UV gel can be a bit thicker or thinner. The thickness is also called "viscosity". The viscosity of the product may increase or decrease by the addition of other additives. For instance, titanium dioxide is a white pigment made of solid particles. Adding a sufficient amount of this white pigment can significantly increase the viscosity of the uncured UV gel. So, it is easy to see that the thickness of the UV gel is determined by the ingredients.

The flexibility of a cured UV gel is NOT determined by the thickness of the uncured UV gel. Flexibility depends on two things, 1. the chemical structure of the oligomers and other ingredients used, and 2. the percentage of cure. Unfortunately, under curing a UV gel will make it seem more flexible. This causes some nail technicians to be fooled into thinking their UV gels are properly cured, when they are not. It is important to note that under curing makes UV gel nail coatings appear to be more flexible. Under curing also causes nail filings and dust to contain larger amounts of unreacted ingredients, which increases the potential for adverse skin reactions such as irritations or allergies. And when these nails are soaked off, these ingredients are released into the solvent and can lead to allergic skin reactions.

Therefore, the consistency of a UV gel is not related to the flexibility of the nail coating. Clearly there is a lot more to know about UV gels and proper curing. For more information about this subject, see Volume I, Topic 15 “The Complexity of UV Curing”.

58:2 A technician in my salon removes UV gel polish by putting her feet in a bag of acetone then proceeded with a Dremel tool. Does the skin on the feet absorb differently than the skin on our hands? Isn't a Dremel primarily meant for woodworking, not salons? Can you explain why a Dremel is not a good idea in comparison to the tools we have specific to our industry?

I agree that submersing the entire foot is not a good idea. I'd recommend using a cotton ball wetted with remover and wrapping it to the toe with tin foil. The entire hand is not immersed, so the entire foot should not be either. Just the fingers and/or toes should be exposed. If done correctly, exposure is low and considered safe.

I don't know that there is much safety difference between an e-file and Dremel. Don’t misunderstand, I do agree that professional tools should be used, but how a tool is used is often more important than which tool is used. Many nail pros use professional e-files incorrectly and frequently damage the nails and/or skin. I suppose a skilled user could use a Dremel tool safely. However, many nail technicians are NOT skilled users, so in general, nail technicians should avoid using Dremel tools. Large numbers of nail techs don't work very safely, tend to over file the nail plate and don't have a good understanding of the natural nail. My opinion is that untrained nail techs should avoid using any kind of e-file or powered tool on their client's natural nails. This is a specialty that requires specialized training or the client's nails are likely to be overly thinned and possibly seriously damaged. A big problem with e-files is that nail techs think that just because they can buy one, means they can use one. They don't bother getting specialized training and some go on to seriously injure clients. Therefore, my message would be, don't use any type of e-file without proper training.

58:4 Can we be allergic to the dust of one gel product and not another? I have been having severe swelling, redness and itching of the skin around my eyes. I have recently been using a new line of gel to try for a few weeks now. I've ruled out all other things I use on my face. This seems to be the only consistent thing. I have no issues anywhere else but my eyes.

No one becomes allergic to a product, they are allergic to an ingredient (or maybe several) in the product. This is important to understand. If that same ingredient(s) is found in another product, that person will develop allergic reactions to both products, because of the common ingredient they share. UV gels are very different from one another, but they often share some similar ingredients, so sensitivity to other products is likely. In this case, I would first suspect the dust. It is too common for nail technicians to under cure their nail enhancements. When this occurs, the dust created during filing will be rich in uncured ingredients. The dust should not come into contact with the skin unless properly cured.

The thinnest skin on our bodies is around the eyes, so I'm not at all surprised that this part of your face is most affected. You should avoid exposure to under cured dust around your eyes. Also, look around a crowded room and I guarantee you'll see several people touching their face, maybe the eye area. We touch our faces more than we realize. Often without thinking, so the face and eye area is often accidentally exposed to monomer liquid or UV gel, etc.

My tips for anyone in this situation are:

1. Make sure you are using the correct UV nail lamp. That's the one specified by the manufacturer of the UV gel and cure as directed.
2. Avoid curing thick layers, the thinner the better, as a rule.
3. Don't cherry-pick and mix different UV gels from unrelated brands/systems.
4. Minimize dust and filings, use an oil designed for use with electric files to keep down dust.
5. Invest in a good quality, professional ventilation system designed for salon use.
6. Avoid touching your face and wash your hands after each client.

Once you become allergic to an ingredient, you will likely be allergic to it for LIFE! Avoid allergies; it's easy if you work safely and correctly. One way is to ensure this is to follow the UV gel manufacturer's directions and heed all warnings on the label. Also, make sure to read the Safety Data Sheets that are supposed to be provided by the seller of the product.

59:2 A UV gel polish company claims their UV gel polish is healthier than others because its pH is 6 compared to the usual pH 3. Is there any validity to that claim?

Nail polish or UV gel polish contain no water, therefore they can have no pH. In order for any cosmetic product to have a pH, a product must contain water. Because no nail polish or UV polish contains water, none can have a pH. This is just marketing puffery. By that I mean, it sounds good, but means very little to the performance of nail coatings in general. Never buy a nail polish or UV nail polish because of its pH.

Now, this is not to be confused with adhesion promoting primers used to improve adhesion by altering the pH of the nail plate. The nail plate does contain water and does have a pH that can be altered, by a primer. I think it's valid to claim that a nail primer can alter the pH of the nail plate to improve adhesion, however, the nail plate will eventually revert back to its normal pH, so the pH effect is not long lasting. That's why other ingredients found in these primers are primarily responsible to create adhesion.

59:5 Can nail primer be used as a bonding layer for UV gel nails? Some use a primer pen on the natural nail, cure it then they apply the building gel. Does the nail primer have to be a specially formulated product?

There are several issues here, but I will try to address them all. First, I don't recommend primer pens, since these are multi-use devices that can become contaminated with nail oils. In my view, it is better to use a brush and bottle to be disposed of after use- not refill. Refilling allows oily contaminants to be transferred from the nail plate to the brush and into the product in the container.

If UV gels require a nail bonder or primer layer which improves adhesion of the nail coating, then it should be used. Otherwise, I would not recommend this to be done. It is easy to see that too little adhesion causes lifting problems, but nail professionals often don't stop to think about the problems caused by TOO MUCH adhesion. Using a primer, when it is not needed can create too much adhesion, which makes removal more difficult, time-consuming and can increase the risks of nail damage related to removal processes.

Improper removal of nail coatings are top causes for nail damage. Many become impatient and use forceful techniques to more quickly remove the nail coatings. When this occurs and nail damage results, many explain away the damage by fooling customers into thinking their nails are just dry, when actually the nail surface has been damaged due to improper removal. My main point is this, if the directions for the UV gel don't "specifically" mention the use of a nail primer, then a primer should NOT be used.

Many times, a nail primer is used as a crutch to hide improper techniques. If the nail technician does a poor job preparing the nail plate, or they incorrectly apply the product, or they improperly cure with the wrong nail lamp, these can all result in lifting- premature loss of adhesion. In these cases, the solution is NOT to use primer, instead, the best solution is to correct the issues that are causing the nail coatings to lift. If the UV gel doesn't require the use of a primer in their directions or instructions, then don't use one or this can create excessive adhesion and the nail coatings will be much more difficult and time consuming to remove. Many nail technicians will forcibly remove the nail coatings- which leads to nail plate thinning, pitting, splits, cracks, peeling and surface white spots. Instead of using a nail primer, it would be wiser to seek out the reason for the low adhesion and solve the problem. Maybe the client's nails need more careful cleaning and preparation. Or perhaps, the professional needs to invest in purchasing and using the correct nail lamp, the one designed for use with the UV gel. Or maybe they should take a manufacturer's class to learn the proper application techniques.

Nail technicians are responsible for their actions and any harm caused by their inappropriate actions. Making up their own procedures and ignoring directions is a common reason for nail problems. Instead of using primers to correct problems created by incorrect use, nail professionals should address and solve the problem correctly.

For instance, nail primer may improve adhesion when nail coatings aren't fully cured, but then nail technicians or clients could develop allergic reactions to improperly cured dusts. The nail technician is not likely to realize they're improperly curing, so they won't take any steps to solve the real problem. Don't make up your own directions/instructions. This is one of the greatest challenges facing the nail industry and a leading reason why most women are afraid of nail salon services and refuse to go to salons. Improper use of nail coatings, is a main reason why the nail industry isn't growing and attracting more customers.

61:3 Lately a lot of nail techs are mixing different nail brands; using one brand of color and another brand of top UV gel. Many say "there is no proof that it's dangerous" and the products are hardened. So why can't we blend them?

Actually, it is the opposite. I would say to those who claim they "can mix nail coating products", where is the proof that this is safe? How do they know this is true? What do they base their information on? Mostly likely, their beliefs are based on lack of understanding about how these products work. Or just on their wishful thinking. There is plenty of proof that these products can cause skin allergies when used improperly. So, for these people to say there is no proof, simply shows they don't know what they are talking about. I'm a scientist who has researched this question for more than 25 years, my opinions are based on scientific testing and fact-based information. There is absolutely no doubt that mixing products as you've described can lead to skin irritation and permanent allergic reactions and this happens all the time! This has been recognized by several government and safety related authorities in several different countries. I don't sell any nail products or nail lamps, so why would I say this if I didn't believe these are the facts? What some don't understand is that each of these product layers would have to be properly cured and this requires exposure to the:

1. correct wavelengths of UV using the
2. proper intensity and

3. for the required length of time.

All three MUST be correct to ensure proper cure. There is no other way to ensure proper cure, and each layer must be properly cured. In other words, each layer would need to be properly cured using the UV nail lamp that was designed to cure the product. A different nail lamp for each layer? That is NOT likely to happen. What these techs don't understand is that UV gels will harden when they are only 50% cured and may only reach 60-70% cure over the next few weeks. When the under cured layer is later filed, both the nail tech and client are exposed to under cured dusts and this can lead to skin irritation and allergies. Many private labelers of UV gels only sell the products and don't understand the science behind these products. That is why some don't understand or teach this information correctly. The biggest risk of developing adverse skin reactions is to nail technicians, since they are exposed daily to these under cured nail coating dusts. I have personally tested the curing of various products under different types of nail lamps. I can tell you that it would not be uncommon for a nail tech to only cure a UV gel to 70%, in fact, it is far too common. Nail techs who expose themselves to these conditions often develop skin allergies. For some it may require six months of prolonged and/or repeated skin contact and for others it may take 10 years of exposure. Sadly, once you are allergic... you will be allergic to those ingredients for the rest of your life.

I get e-mails from nail techs who are desperate to save their careers and want to know what they can do to stop their skin from overreacting to the products. The facts are, sooner or later many nail techs and/or clients will develop permanent allergic reactions because of improper cure. I hear from many of the nail techs to which this has happened. They're usually very embarrassed and usually don't want me to use their names on social media. Sometimes they don't want other nail techs to know about their issues because they are leading nail educators who improperly cured and mixed products for years, and now they have skin allergies that worsen with each exposure. Many nail technicians are forced to get out of the nail industry and can no longer work with nail products. Here's a hint: Take a look at the hands of some veteran nail trainers who teach these risky methods and you'll find many have product sensitivities, which is not a coincidence. Don't copy their bad habits or repeat their same mistakes.

63:1 "I know there are several medications that cause a heightened skin sensitivity to UV exposure. Is this the case for nails as well?"

Yes, if you are using a medication that heightens sensitivity to UV, you could have an adverse skin reaction with a UV nail lamp. The nail plates themselves are not adversely affected, but the skin may be more easily burned while using these medications or treatments. This is called a "Photosensitivity" and this can lead to skin irritation or an allergic response. This can occur with medications taken orally or applied topically. The most well-known example is the antibiotic tetracycline, but there are several dozen medications that can cause these types of sensitivities. Typically, the skin will react within minutes, but sometimes the reaction is delayed a few hours. Often the reaction will appear as exaggerated redness or swelling and blisters can form. The skin can become very itchy. The reaction isn't always isolated to the area of exposure and even adjacent unexposed skin can display symptoms. Wearing sun screen or covering the hand with a cloth or UV protective gloves can help minimize exposure. However, for advice that involves medication, it is always best to ask the prescribing physician. In these cases, the doctor will most likely advise their patient to skip such services, until they've completed taking their medication. Better to be safe than sorry, as the saying goes.

63:4 I have a long-term client with no issues apart from some surface peeling. Now suddenly, the nail plates are badly stained, after wearing a dark red color. I'm properly curing with the correct UV nail lamp, I don't think I'm applying the color layer too thickly either and I use a thin coat of base.

One of the first things to consider is the base coat, since applying it too thinly can increase staining. Red colored nail coatings are one of the greatest challenges for the nail plate. It is difficult to create a deep red color that doesn't stain the nail plate. Because some red colorant molecules can penetrate the nail plate's surface. When that occurs, the molecules collect and pool just underneath the surface to create a visible stain. The surface of the nail coating may seem hard, but if the color coat isn't completely cured all the way through, it will remain tacky in the area where it meets the base coat. This is called a "tacky interface", in technical terms. This tacky interface allows colorants to move around more easily, into and through the base coat, so nail plate surface staining becomes easier.

This is especially true for areas of nail surface damage. Nail plates that are over filed, peeling, pitted or otherwise damaged are more readily stained. This is hard to avoid when nail surfaces are damaged. The greater the damage, the more likely it is that staining will occur. Also, it is very important to note that staining is more likely to occur when the color coating is applied too thickly. So, besides finding a solution to the surface peeling problems, perhaps what this nail professional needs is more base coat and less color coat.

63:5 Unfortunately, my initial training using UV gels wasn't great and I had overexposed myself before I knew it was a problem. Now, despite taking plenty of precautions I have severe reactions. I am very careful to use nitrile gloves and protect my hands, but sadly it's not working. I have the most severe case of allergic dermatitis that is debilitating at times. I have had allergy

patch testing and know I am allergic to my products. Is there any hope for me? Or do I have to start considering giving up my job?

Wow, what can I say? Other than “Don’t let this happen to you”. The same will happen to many others because they don’t take the necessary precautions to work safely. People don’t become allergic to products, they become allergic to ingredients, so switching products usually won’t help them. Often these problems are a result of mixing products not intended to be used together or using the wrong nail lamp OR using pigments, glitters or other colorants that are not safe for cosmetic use. Or just from constantly exposing the skin to under cured dust/filings. Sadly, the best answer when extreme conditions like this occur is, “find a new line of work and stop using or wearing nail coating products forever”. Once a serious allergy develops, it will most likely worsen with each additional exposure. It sounds like that’s what has happened to this soon to be “former” nail technician. At the first signs of skin problems, this nail technician could have prevented problems from worsening simply by ensuring proper cure and avoiding skin contact with dust, filings, sticky layers, UV gel, etc.

Keeping brush handles, containers and other objects free of sticky UV gel residues is also important. To make gloves more effective, some will wear “barrier creams” underneath their gloves and on the wrist and forearms. Barrier creams are skin coating creams that can slow penetration of allergy causing substances into the skin where they activate allergic reactions. However, barrier creams are NOT replacements for gloves and they must be used cautiously, since they usually contain ingredients from the silicone family. Even small traces of silicones can block adhesion if they contaminate the nail plate, so be sure to carefully clean the client’s nail plates and avoid getting any silicone containing cream or lotion on tools or brushes. Finally, protect your career and the investment you’ve made to get where you are today. Please be serious about working safely.

Image 3: Disposable nitrile gloves are the best choice for salons, 8 mil glove thickness is best.

65:1 "Is heating up of the UV gel a problem for other nail techs, or is it just me?

This is a problem that occurs far too often and should be avoided. When any chemical reaction releases heat, it is called an "exothermic reaction" or "exotherm". An exotherm can be very significant and even damaging. I hear about this problem all the time. It was always my chief concern when developing any artificial nail coating, because they can happen with any and all artificial nail products- and I know of no exceptions. It's the nature of the chemistry involved, and it is unavoidable. Even so, exotherms can be minimized and controlled. One way to control exotherms is to ensure the product is carefully and responsibly formulated by someone skilled in the art of their development. I went out of my way to minimize exotherms whenever I developed any nail coating. This was one of my top goals for product development. Exotherm is a real and significant problem. I even spent months building a special scientific instrument that would very precisely measure the exotherms created on the nail plate during curing.

Exotherms can be a red flag when a product over heats to the point of causing burns to the nail bed. They can lead to onycholysis, which is a medical term for the separation of the nail plate from the nail bed. Excessive heat can be a sign that something is seriously wrong and should be immediately corrected before the client develops a permanent injury, which can lead to infections and/or loss of the nail plate. Exotherms are made to be much more noticeable and painful when a nail tech friction burns the nail bed from overly aggressive filing, which magnifies the problem by making the nail bed more sensitive to heat. What other factors can cause exotherms? As mentioned before, using the incorrect nail lamp to cure a UV gel is a leading cause. LED nail lamps emit more UV than traditional nail lamps, so when used with a nail coating not designed for use with LED nail lamps, this can lead to significant over-heating and nail bed burns that can lead to onycholysis. The thicker the applied coating, the more heat it releases during cure. Another common reason has nothing to do with the formula, but everything to do with over filing. When the nail plate is filed too aggressively or with a heavy hand, this can friction burn the nail bed and make it more sensitive. When the nail bed is injured by over filing, it becomes much more sensitive to heat, making it more noticeable. The points raised by this question are likely reasons for excessive heating, so these are the first places you should look when trying to prevent this potentially harmful condition.

67:5 An educator in Brazil says that LED gel is curing from the bottom up and is therefore more resistant to breakdown, while the UV gel is curing top down and are less resistant. I never heard that. This is true?

This is not correct, and it makes no scientific sense. All UV gels cure in the same fashion, regardless of the UV source. LEDs and fluorescent tubes are two different types of UV sources. UV gels contain various curing agents that depend on the intensity of the wavelengths and the length of exposure. LED-style nail lamps emit UV with greater intensity than traditional fluorescent-style UV nail lamps. Even so, no type of UV gel cures from the bottom up.

67:6 A friend has an allergy that she believes is caused by uncured gel on the nail plate entering the body through this route, but I suspect it is from over exposure to the UV gel or dust. Aren't the molecules of ingredients too large to enter the nail plate?

You are correct, it would be extremely unlikely that penetration through the nail plate is the cause of the allergy. The exception to this would be if the nail plate were highly damaged or missing portions that expose bare nail bed. If the bare nail bed is exposed to UV gel, the risks of developing an irritation or permanent skin allergy will significantly increase. UV gels should only be applied to an intact nail plate. The skin has its own immune system that is separate from the internal immune system. This explains why direct skin contact is required to cause skin allergies. UV gel ingredients do NOT penetrate through healthy, intact nail plates to cause allergy, instead, this is from direct exposure to UV gel, dust, filings or inhibition layer. This becomes especially more likely if the UV gels are not properly cured.

68:5 If I use an LED nail lamp on a product designed for traditional UV nail lamps, will the UV gel become over cured?

Yes, it can lead to over curing and this can cause a nail coating to overheat, burn the nail bed, and lead to separation of the plate from the bed. Bacteria or fungi can then infect these areas more easily. UV gels should only be cured using the UV gel lamp or lamps specified by the UV gel manufacturer. LED style nail lamps emit much more UV than fluorescent-style UV nail lamps and these levels may be too much for traditional nail coating products, unless they are specifically designed for LED lamps. If a UV nail lamp was not specifically designed for the UV gel being used, over curing is a very real issue that can lead to injury and harm, so I would warn against doing this. Many things can go wrong when nail professionals don't use a nail lamp that properly cures their UV curing products. That's why I recommend only using the nail lamp that is specified by the manufacturer of the UV product. Note: it doesn't matter if a company makes the nail lamp especially for the UV gel, what matters is if they UV gel is formulated specifically to work with one (or two) lamps. No UV gel can be formulated to properly cure with any nail lamp, no matter what type of nail lamp it is. Some mistakenly think that so-called "dual lamps" will properly cure all types of UV gels, but that is a false assumption. As I've pointed out in other questions, both the wavelength range AND the intensity of the wavelengths are very important to ensuring proper cure and no one lamp can properly cure all UV gels.

68:8 Why are UV gels thick and can anything be added to thin them out?

No, nothing should ever be added to UV gels to make them thinner, because this can disrupt the chemistry and adversely affect cure. The result could be an increased risk of developing an adverse skin reaction, including the potential for permanent allergic reactions. Never add solvents or any type of thinner to a UV gel unless instructed to do so by the manufacturer's directions or improper cure becomes a greater possibility. If instructed by the manufacturer to do so, precisely follow those directions or problems can occur. The base ingredients are called "oligomers" which are molecules that have been partially pre-polymerized into short chains. These short chains are more prone to be tangled, than small molecules, which makes the product feel thicker or more "viscous". Then when UV gels are exposed to UV energy, these short chains join to create the very long chains that solidify to create the final polymer coating. This is done so the UV gel can cure more quickly than it normally would. Very much like precooking food so it can be more quickly finished in the microwave. The oligomers are the reason these products are thicker than other types of nail coatings and adding solvents to thin them will result in the creation of a significantly weaker nail coating.

69:6 Wondering about your thoughts on embedding REAL shed snake skin into gel nails? This seems horrifying to me. Snakes carry salmonella. Can the skin be disinfected properly? I'm seeing nail educators saying it can. Will you "shed" some light on this topic?

Even if the shed snake skin doesn't contain salmonella, it should be properly cleaned and disinfected before using on a client's nails. If this is done, the risks are probably negligible. However, there is a bigger issue that should be considered. Embedding anything in a UV gel can block UV from penetrating and this can lead to improper cure of the lower layers and lead to under-curing. Why? Anything that blocks UV penetration through the nail coating can adversely affect the cure. If the UV doesn't penetrate through the layers of UV gel, improper curing becomes much more likely. Improper curing is a leading cause of skin allergies and too many nail techs are regularly exposed to improperly cured UV gel. I know it is fun to embed items into UV gels, but this will very likely lower the degree of curing and could increase the risks of the nail tech developing skin sensitivities and permanent allergies, if they are repeatedly exposed to under cured dust/filings. Beware of improperly cured dust and filings. The best way to avoid them is to ensure that nail coatings are properly cured.

71:6 Can the impact of hormones change the level of adhesion of UV gel to the natural nail plate? Especially the menstrual cycle?

It is possible for nail plates to be affected by hormonal changes occurring over long intervals, which rules out the effects of monthly menstrual cycles. I've never seen any evidence or other indications that menstrual cycles can affect adhesion, nor do I see how they could, so I'd label this as a myth. During pregnancy, hormones are released into the body and they can cause the nail plate to accelerate growth, so it is clear hormones can affect nail growth. However, the nail coating is adhering to nail cells that developed in the nail matrix prior to product application. Hormones can affect newly forming nail cells, but it is pretty unlikely that hormones can affect them months later after they grow out and are just as unlikely to cause poor adhesion.

For the same reasons, it makes no sense that menstrual cycles would affect adhesion. This would be akin to saying menstrual cycles cause split ends in hair, but how could they affect hair that has already grown out? This would be blaming spilt ends on the wrong things and would never lead to a proper solution that prevented spilt ends. The same is true for claiming that hormones affect nail adhesion- it's mostly used as an excuse to shift the blame and will NOT lead to a solution for problem. I recommend that you should first look at your own work and then the client's lifestyle and hand-related occupational exposure as the step toward finding a solution to adhesion problems. That will be the best way to solve such issues.

72:1 Can winter temperature cause UV gel to lift from the nail plate, especially on customers who have very brittle and weak natural nails? They don't have lifting when I use a flexible builder gel, but now temperature is below zero and many customers are having problems.

This nail professional is correct, nail coatings will lose flexibility and become more rigid and will crack or break more easily when cold. The colder the temperatures, the greater the negative effects.

In part, this occurs because toughness and durability are significantly lower as temperatures drop. The nail coating could eventually lose its flexibility and become rigid, even brittle. UV builder gels with higher flexibility also lose some flexibility, but being already highly flexible, these can retain some flexibility even in the coldest temperature. The result is- more flexible nails will remain tougher and more durable at lower temperatures.

Adhesion can also be affected by temperature. Here is how that works. Everything expands or shrinks when heated or cooled and the greater the change in temperature- the greater the shrinkage or expansion will be. Interestingly, different materials will expand or shrink to different degrees and at different rates. These variations depend on the chemical composition and molecular structure of the substance. As a result, the nail plate does NOT shrink at the same rate or to the same degree as the nail coatings. Because they shrink at different rates, this creates a mismatch that builds stress at the interface between the nail plate and the nail coating. This stress can pull them apart and will negatively affect adhesion. The reverse is true when these warm up again; one will expand faster than the other and this also causes stress to build up. As the nail plate and nail coating expand at different rates, stress at the junction where they meet will increase. Each time this occurs, there is weakening of the bond between the nail plate and nail coating. This is true for all nail coatings. When flexibility is lost at the same time the nail coating is shrinking, the forces generated may seem tiny, yet they're enough to cause lifting. Tell your clients to protect their nails and always put on warm gloves before going outside and don't freeze their nails. Be especially careful to avoid rapid changes in temperature, since this can cause even greater problems that will cause slower, more gradual changes.

72:4 Is it safe to put your hands / nails into a nail lamp after you're diagnosed with malignant melanoma skin cancer?

Clients diagnosed with skin cancer have a medical condition and their doctor should determine if this is safe or hazardous for their patient. The doctor's decision would be based on the patients' medical history. I do NOT think this is a decision that a nail professional should ever make, because it is a medical choice, not cosmetic. What if the "patient" wants to be a nail client and insists that you do the services? That's a little more difficult to answer. Please understand this, a client can NOT sign away or give up their right to have a safe service. In other words, a client can't give you permission to do something that goes against your professional judgement, if you believe it would be harmful or highly risky or potentially dangerous. The customer is NOT always right. As a professional, you are always responsible for working safely and protecting the client's health. So, if an unhealthy condition exists, you should refuse the service. My guess is that most doctors, out of an abundance of caution, would likely advise against exposure to UV nail lamps. But some may recommend protecting the hand with a cover or SPF lotion. Or a disposable UV hand shield/glove can be extremely effective in blocking virtually all UV exposure. That's why it's always very important to get the doctor involved when clients are making such choices. Don't go around the doctor, remember, you're working on their patient- it's not just your client any longer.

72:5 Is there such a thing as a protective glove that's a good UV shield for hands? Do you get that from a doctor?

Yes, there are and you don't get them from a doctor. These shields can be extremely effective at blocking UV. Fortunately, the skin on the back of the hand is the most UV resistant skin on the body, so we've got that going for us. UV nail lamps have been carefully tested and shown to be safe and not at all likely to cause skin cancer, but many clients are STILL concerned about photoaging to the backs of their hands. They could apply high SPF lotion to the back of the hands before the service and wait until it is dry or they can use a less messy alternative- disposable UV protective shields, which can be purchased on the Internet.

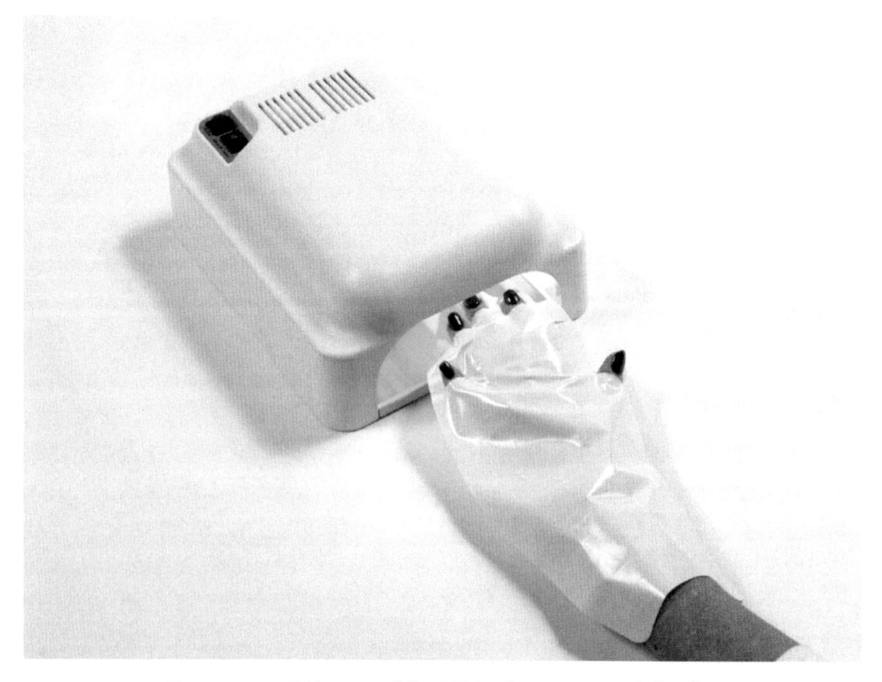

Image 4: Disposable UV gloves provide the ultimate UV protection
(Photo credit Renee Albera, youveeshield.com)

Here's one example of a disposable UV shield that I am aware of, however there may be other solutions for blocking UV that are available. http://www.youveeshield.com/ If premature aging is a concern, these shields lower UV exposure to virtually non-existent levels. They don't cover the nail bed itself, no worries there as the nail plate has a high natural SPF so the nail bed is naturally protected against any UV exposure from nail lamps. Also, the nail matrix is covered in a thick layer of skin, so overexposure to the matrix is very unlikely.

73:1 I saw a picture circulating on the Internet showing a dark stripe running the length of the nail plate that was supposed to be nail cancer caused by a UV nail lamp. Is this possible?

I saw the picture as well and was asked this question by nail technicians in seven countries. The dark stripe was about the width of a drinking straw. The message attached to this image incorrectly identified this stripe as "nail cancer" caused by UV nail lamps. It was posted by someone claiming to be a friend of the person with the nail stripe, but no other information was provided. The image and message was repeated widely on the Internet, even though there was no information to back up this silly and highly speculative claim. Even so, many websites pounced on this as if it were fact, when actually this is a prime example of fake news and shows how fake news can unfairly harm your business. Personally, I am convinced this particular piece of "fake information" was planted by fear-based activists for purposes of misleading the public about UV nail lamps. Yes, I've seen this done many times before by these fear-based activists. This was a crazy claim, with NO supporting evidence, but evidence isn't needed if the goal is to create irrational fear. Here are the facts, dark streaks like this are a relatively common condition that occurs on hands and feet of men, women and children who have never been to a nail salon so it is NOT caused by UV lamps and there is no reason to believe this can occur.

These dark stripes are very common around the world, mostly to those with dark skin and is relatively uncommon with in those with pale skin. Many things can lead to the occurrence of these dark stripes, they happen when the cells in the nail matrix begin to produce melanin. This can be a result of pregnancy, nail biting, carpal tunnel syndrome, psoriasis, viral warts, hyperthyroidism, exposure to x-rays, phototherapy, medications, excessive iron in the diet, and many other things. In rare cases this could indicate the existence of a tumor, but these are FAR more likely caused by recreational sun exposure- not UV nail lamps.

The nail plate has a high natural SPF that protects the nail bed and the matrix is buried under a thick layer of skin, so it's VERY unlikely that UV nail lamps would cause such issues. These UV lamps have been repeatedly demonstrated to be safe as used. In short, there is no reason to believe this story no reason to believe this dark stripe was caused by an UV nail lamp, and there is NO reason to believe the person who posted this information. Especially since she refuses to return my e-mails or communicate with anyone else who tried to contact her and follow up on these claims.

Personally, I suspect this is an entirely made up story, which would mean those behind this nonsense were INTENTIONALLY trying to harm the business of salons everywhere. That's becoming commonplace. Fear-based activists regularly attack the nail industry with misinformation to sow fear and they often dupe the news media into spreading their trickery. That's why it is so important that nail professionals become more educated about the nail and nail products, so they can fight back against these tricky deceptions. Don't be fooled, get the facts from someone who knows. That's why I do videos and write Face-to-Face with Doug Schoon books so that I can provide nail professionals with the facts! Tired of silly Internet myths? Then arm yourself with the facts and knowledge you need to see the truth. You can find out more about my video series at: www.FacetoFacewithDougSchoon.com.

Polish, Other Coatings and Adhesives

54:2 I was told that a product containing cyanoacrylate is a "gel", but this component is a glue. How can it be both? I think it can only be one or the other.

Cyanoacrylates are monomers from the acrylic family that are used for many purposes in the beauty industry. For instance, they are used as tip adhesives for nails, used to adhere rhinestones and other accessories to nails, they're also used to create fiberglass and silk wraps, and they are regularly used as eyelash adhesives. They are also sold as "no-light nail gels" and for so-called "dip" systems.

In all cases they are considered adhesives, since they stick to the nail plate and more commonly they are called "glue". Cyanoacrylates can be thin and watery or thickened to a gel-like consistency. Some mistake the term "gel" to mean only UV curing gels. UV curing gels are called "gels" because of their gel-like consistency. In other words, not all gels are UV curing. The word "gel" describes the consistency of the product, not its chemistry or composition.

Hair gels for instance are thickened to a high viscosity, which is why they're called gels. Therefore, it is entirely proper to refer to a thickened cyanoacrylate as a gel. Nail professionals should understand that not all gels are UV curing gels. That's why when referring to UV gels, they should not just be called "gels"- which is an overly simplistic name. This is a great example of why it is important to use more descriptive terms when talking about nail products, if for no other reason, to reduce confusion. Another example of this are the sophisticated, high tech monomer blends used to create artificial nail products. These products are among the most scientifically advanced products in the beauty industry, yet nail professionals refer to them as "liquid". The polymer powders are equally sophisticated and complex, yet they are simply called "powders". If nail techs want clients to recognize them as professionals, they would be wise to use more professional terms, such as UV gel, monomers and polymers.

54:3 I read in your book that nothing you can apply to the nails will make them grow faster or stronger, since that is determined by the nail matrix. But some nail products claim to fuse together the nail plate layers to protect the nail and prevent cracking and breaking and state that when they are used, the nail plate is stronger and grows longer. Is this a false claim? How could they possibly work?

I hear this question a lot and it is a constant source of confusion. Cosmetic products can NOT legally claim to make the nail plate grow faster, but that is not what these products are claiming to do. When you carefully read the claims, and take the time to understand what is really being said, this becomes clear. These products don't claim to make the nail grow faster. But they do claim to make the nail stronger. These products reinforce the nail plates making them stronger which prevents them from breaking, so they can naturally grow longer. Grow is the key word here. Cosmetics can't and don't affect the way the nail plate grows. But they can reinforce the existing nail plate to make it stronger, so it can naturally grow longer. That is hugely different than saying the product makes the nail grow faster, which would be a false claim. Instead, these products claim to make the nail plate stronger, so it won't break and can therefore naturally grow longer. I think this is a valid claim. It pays to carefully read marketing claims to make sure you properly understand what's being said and if you don't understand, I suggest contacting the manufacturers to ask for a better explanation that you can understand.

57:2 I remove my client's nail coatings with pure acetone, using the foil and cotton method. After about 5 minutes some of their nails will start to curl inward. Why and what can I do to prevent this.

I would guess that the nail plates which are curling are thinner than the other nails. The acetone will absorb some water which will lower the water content of the nail plate. The nail plate normally contains about 15% water. Even when you lower the water content even by just a few percent, it can make a big difference, especially for thin nail plates. Much of the water inside the nail plate is located between the layers of nail cells. The science is a bit unclear about how many layers make up a "typical" normal nail plate, but the latest research suggests that about 50 layers of nail cells is a good estimate. Whenever the water (aka moisture) leaves the nail plate, these layers will draw closer together. This can slightly alter the shape of any nail plate, to varying degrees. Thinner nail plates are much more likely to curl when they lose a little water.

The first thing I would try is to regularly perform hot oil manicures, once or twice per month and daily use of a high quality, penetrating nail oil. Doing so will increase the oil content of the nail plate, making it more difficult for water to leave the nail plate's layers, and therefore slow down or prevent thinner nails from curling. Also, avoid filing the nails even thinner than they already are. This is one of the biggest mistakes nail technicians make, over filing the top surface of the nail plate. This out-of-date technique is still taught by many nail schools and training courses. Respect the nail plate. The less this upper surface is filed, the better. It's better to keep the nail plates thick and strong!

58:1 I've been told that nail plate yellowing is due to nitrocellulose and not pigment stains. Recently, I used a different base/top coat and the client's nails after the removal of a light lavender color nail polish were very yellow. Normally my clients don't have this type of yellowing no matter of the color of the polish they are wearing.

Pigments are not likely to stain the nail plate. They are too large to penetrate. Dyes and Lakes are other types of colorants that are often used, and many can and DO stain the nail plate. There are three different reds and one yellow colorant that have been reported as the most likely to stain the nail plate. The reds colorants are listed on product ingredient label as Red no. 6, Red no. 7, or Red no. 34. In the European Union, all three of these red colorants would be sold under their color index number "15850". The yellow that is reported to cause a lot of staining is Yellow #5 Lake, which in the European Union is labelled as "19140".

It is true that some grades of nitrocellulose can stain as well. But the more expensive, higher quality grades are much less likely to discolor. Smaller companies that don't sell much nail polish often use these inferior grades, since the best grades are too expensive when purchased in low quantities and the larger companies buy up all the high-quality nitrocellulose. When it does stain, nitrocellulose tends to be a more brownish-yellow stain, while discoloration caused by colorants tend to be the light yellowish tones or other odd shades, such as green.

If this base coat contained nitrocellulose and no colorants, the nitrocellulose is a possible suspect. However, a high-quality base coat should not stain the nail plate, unless its surface is damaged. Damaged nail plates are much more likely to pick up stains, even from some foods stains or clothing. Damaged nail plates will absorb stains more easily than healthy nail plates. To say the nitrocellulose is responsible for nail yellowing is an over simplification. Marketers tend to oversimplify things for three reasons,

1. They don't understand the issues, so instead of giving the facts they just keep it simple to hide their lack of understanding.

2. Many nail techs often don't understand these issues, so the message must be kept simple.

3. Marketers focus on only what helps them sell their products and often exaggerate the information.

It is definitely an over simplification to say that nitrocellulose is what stains the nail plate. More likely the causes are "dyes" and "lakes".

59:3 I was told that nail glues and fiberglass resins can cure either by air or by moisture? Is that true?

These glues and resins are made with monomers from the acrylic family. They are called cyanoacrylates. These products do NOT cure by air and instead are cured by moisture. Air is a complex blend of many gases, including mostly hydrogen and a smaller amounts of oxygen. These gases have no ability to cure cyanoacrylates, besides "air" is found inside the product container after they are filled, shipped and stored. Like paper or fabric, many things absorb some moisture, including air. When the air absorbs water we call this "humidity".

The higher the RH or "relative humidity", the more moisture there will be in the air. Relative humidity tells us how fully saturated the air is with water, so 100% RH would be completely saturated air, otherwise known as "Hey, it's raining". Both liquid water and water vapors cause cyanoacrylates to cure. Water vapors exist in the air and they can accelerate the cure of cyanoacrylates to harden and cure to create polymers. The same is true for the nail plate which typically contains about 15% water.

Cyanoacrylates are stored in water-free and acidic conditions. Water is a weak neutralizing base which counter acts the acidity and shifts cyanoacrylates to a basic condition with a pH slightly higher than 7. This triggers polymerization. Usually RH's effect on set time isn't noticed, unless the relative humidity changes drastically. Relative humidity below 40% can slow set times, while above 60% RH, set times can hasten hardening. Set or fixing time is determined by two main things.

1. How much moisture is present on the surfaces being adhered together.

2. The temperature of these surfaces, which is the temperature of the nail and the product being applied.

Catalysts can also determine set time. They are substances which dramatically increase set times, as well, but without moisture, these catalysts could not do their job. So it should make sense that nail plates with a higher moisture content will cure cyanoacrylates more quickly than those with a lower moisture content. Nail plates with 20-25% water content will cure cyanoacrylates more quickly than plates that contains a low amounts of moisture, such as 10-12%.

Many things on the basic side of the pH scale can also trigger cyanoacrylate polymerization. Cotton towels, depending on the detergent they are washed in can have an alkaline pH. This explains why cyanoacrylates can instantly harden and produce fumes when dropped onto a towel, since the pH is changed from slightly acid to slightly alkaline (aka basic).

The pH Scale

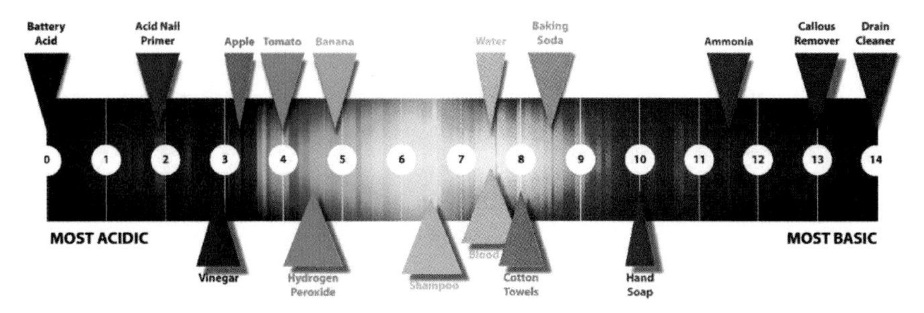

Image 5: pH scale and the pH of some common items.

60:1 Is it necessary to wait 24 hours to apply UV Top coat or the nail polish underneath won't dry properly under the UV gel topcoat and would be unsafe?

I am happy to hear you are concerned about being safe, but I can assure you this issue isn't a matter of safety. Instead, this is about performance. If the performance of the nail coating is compromised, it won't meet the customer's needs or expectations.

The longevity of the nail polish can depend on how much solvent was trapped under the UV gel after it was hardened. An increased amount of trapped solvent can result in a greater chance for some type of service breakdown. For this reason, it is best to wait until the nail polish has at least hardened on the surface. The majority of the solvents will escape from nail polish in the first three minutes, so that is probably long enough to wait before applying a UV gel top coat. The remaining solvents will not be permanently trapped and will slowly escape via other routes near the side walls. However, always consult the manufacturer's directions for special instructions. Why is this NOT a safety issue? Studies have shown that the total amount of solvent exposure for clients and nail professionals is very low and well within safe levels. That's why I feel confident telling you that residual solvent exposure is not a concern as long as application is done in a well-ventilated area, as indicated in the instructions. Some have exaggerated this issue into more than it is and are creating unwarranted fear. Most of what you hear about nail polish being unsafe is silly nonsense invented by a fear-based activists who use this deception to trick people into donating money to them. These groups distort the facts, simply to get people to donate money to them, so don't trust them. Nail polish has been considered safe for more than 80 years, so there is little reason for concern.

60:4 Is nail glue or silk considered "vegan"? Are there any rules about making this claim?

There are several different definitions for "vegan", depending on who you speak to about the subject. Most think of "vegan" as being related to food and these are not foods. They would say the vegan claim does not apply and it is wrong to make these claims. Nail glue contains no animal by-products and never has, so they would say that all nail glue is vegan. However, in my view that would be a deceptive claim since this wrongly infers that other nail glues do contain animal by-products and that is false. This would be a bit like selling bottled "vegan water". All bottled water can make the same claim, but don't because it is a ridiculous claim.

Silk is a bit different. Silk comes from a silk worm, so it is made by animals, but it is not a part of an animal. It's more like a bird nest, made for short term use. Since this is cast off by the worm, it isn't a part of the worm, therefore some consider this to be "vegan". Others would vehemently argue that even silk is an animal by-product. Try asking ten people and see how many different answers you get. So, it really comes down to your own definition. Companies can say what they want, since I am unaware of any regulations that dictate how this term must be used. My experience is that often, companies claiming to be vegan, are often no more vegan than their competitors who don't make the claim. I see "vegan" as being more about what you eat, but others take it further and would disagree. In many cases, being "vegan", it seems is often more a matter of opinion.

60:5 What about nail glues? Are these tested on animals?

Cyanoacrylates are the monomers used to create nail glues. Cyanoacrylate monomers are used in surgical medical procedures which require extensive and rigorous safety testing that includes some animal testing. Therefore, it is fair to say that the medical profession at some time in the past tested cyanoacrylates on animals. They are forced to do this because safety substantiation for medical devices is legally required by most, if not all, countries. However, the cosmetic industry does NOT test these glues on animals. The safety of these adhesives was established long ago, so there is no need to prove they are safe by these means. In fact, it is safe to say that animal testing is NOT an issue for the nail industry and I know of no nail companies that perform such testing.

65:4 I'm getting a lot of questions from clients about the safety of peel-off basecoats. Are they ok as long as people peel from the base of the nail plate toward the tips to go with the grain of the keratin cells?

If they did as you described, then they would be going "against the grain", since any so-called grain in the nail would run across the width of the nail plate. Keratin fibers inside the nail cells lays across the width of the nail plate, which is why the nail plate typically splits across the width of the plate and not down the length. Therefore, the grain doesn't seem to make a difference, at least when it comes to nail surface damage. People with healthy nail plates may not see damage when these base coats are used according to directions. Even so, if the nail plate is already damaged, these peelable nail coatings can worsen existing damage. If the nail plates become dry-appearing or small white spots develop on the surface of the plate, then I would recommend that the wearer should discontinue use and/or revaluate how they are using and removing these products.

These types of peel-off basecoats are more likely to worsen existing nail surface damage, but it is very common that people ignore directions and do it "their way" then blame the product when it doesn't work. Users will likely notice decreased wear time. In other words, they won't get the same wear. These basecoats are likely to shorten wear time by as much as 50%. Often those who use these types of products are people who want to change their nail color often and doing so can also increase nail damage.

66:4 As a practicing Muslim, I am only allowed to wear certain types of nail polish, those which adhere to Islamic law. Do you know which ones are allowable?

To be clear, I am not Muslim, and I do NOT have the qualifications needed to interpret Islamic law, but over the last several years I've consulted with several Muslim experts in these areas and I have relied heavily on their expertise. By combining my scientific knowledge with their expert knowledge about specific Islamic laws, I've been better able to determine the facts behind this highly confusing issue, as it applies to wearing nail polish. Many companies have recently started marketing "Halal" nail polish to Muslim women, which has led to confusion about whether these nail polishes adhere to Islamic law. Most of the confusion exists because the term "Halal" implies that all Islamic requirements have been met, but that is not the case for nail polish. Nail polishes have a separate and different requirement that goes beyond being Halal.

First, let's discuss what it means to be "Halal certified". What is that? Halal certified- simply means "permitted or lawful", under Islamic law. This certification is provided for foods or cosmetics if they can demonstrate that they do not contain any substances forbidden by Islamic law, e.g. drinking alcohol which is called ethanol and pork are typical examples. Halal certifications are intended for foods, but also include other items that may be swallowed, e.g. mouthwash, toothpaste or lipsticks. This requirement really doesn't apply to nail polish, yet some marketers are using the term in a confusing fashion and some are misleading many Muslim women. Halal certification is very easy to obtain. Companies must pay a monetary fee to the governing body, after which they review the list of ingredients for the product.

Once it is determined that a food or cosmetic product contains no forbidden substances, it can become "Halal certified". This is very much like the Kosher designation. Both Kosher and Halal have similar goals and requirements. However, nail polish is different from other cosmetics and MUST adhere to another Islamic law that so far has proven impossible for any nail polish to achieve. This law requires Muslims to wet their nail plates before praying, and this can NOT be achieved if they are wearing nail polish. A "spiritual wash" called Wudu, is required to be performed before every prayer and this is extremely important. This ritual wash is used to attain a state of mental preparation for prayer and help to maintain good physical cleanliness. Muslim doctrine teaches that practitioners MUST perform a spiritual wash BEFORE they pray. If they do not perform this required spiritual wash, their prayers are considered invalid and do not count. In other words, praying without properly performing Wudu, is like not praying at all. Pretty serious stuff for practicing Muslims.

Since Muslims are required to pray five times per day, they must perform the spiritual wash several times per day, which is why most practicing Muslim's don't wear nail polish. If a Muslim does not get their nail plates wet with liquid water, their Wudu was not properly performed and thus their prayers do not count. Claiming that water "vapors" penetrate through nail polish to the nail plate is not nearly enough; vapors and traces of moisture do not count as a spiritual wash. I have examined the ingredients and the testing done on nail polish that claims to allow water to pass through the polish to reach the nail plate. In my opinion, these are nothing more than simple tricks designed to fool the public. I'm speaking specifically about a silly test where nail polish was applied to a coffee filter and then it was shown how water passed through the filter. What? Since when did a coffee filter become a good substitute for the nail plate? This is utter foolishness and trickery, in my view. I suspect that I could make any nail polish pass this silly test, so it proves nothing. I believe this test was adopted to fool people into thinking they could perform a proper spiritual wash without removing their nail polish, but that is a false hope.

As a scientist with more than 25 years' experience developing nail coating products, I can assure everyone that there are NO known formulations of any nail polish that allow water to pass through the coating to fully wet the nail plate as required. Here is the simple truth is; no nail polish is compliant with the requirements of Wudu spiritual washing. Nail polishes that claim to be "Halal" compliant, are simply declaring they contain no forbidden ingredients and that has nothing to do with whether nail polish is Wudu compliant, as is required. Halal cosmetics are fine to wear, just as long as they do NOT interfere with Wudu spiritual washes before prayer. Nail polish does interfere with Wudu, therefore practicing Muslim women who care about water reaching their nail plates should be advised that they will still need to remove their nail polish before performing a spiritual wash and prayers. Marketers should be careful not to over promise what their products can do and should be clear about their limitations.

After 25+ years researching the natural nail, I have a deep understanding of nail coating products and their chemistry, so I'm often asked, "what about the future"? Will there ever be a fully compliant nail polish that Muslim women can wear continuously and still pray regularly? I do believe that someday, a fully complaint nail polish that allows water to pass through the coating to wet the nail plate will someday be developed. To make such a huge technological breakthrough will require a significant research effort, which now-a-days most cosmetic companies try to avoid. Only the biggest and the best can take on a significant challenge like this, so it's not likely to happen any time soon and not likely to be developed by a small company. To date, all the money has been spent on marketing of "Halal" certifications and no one has been willing to invest in a real research program to create a fully complainant nail polish. But with approximately 500 million practicing Muslim women of nail polish wearing age, I suspect that sooner or later someone will create a useful and viable solution to great unfilled need.

70: 4 I know cyanoacrylates are called by the name resins and adhesives, but are they also classified as monomers?

Yes, the term "monomer" is often misunderstood. A monomer is any molecule that can join to create a polymer. Many in the nail industry misunderstand this word and mistakenly believe the "monomer" is the name of the acrylic liquids used to create artificial nail coatings, but that is an overly simplistic view. Monomer is a generic term for ANYTHING that can polymerize. Any molecule that can chemically join to create very long chains are called monomers. Sugar molecules like "Glucose" are also monomers that can be polymerized into a polymer chains. Amino acids are also naturally occurring monomers that are joined into long chains to create polymers called protein.

74:3 My nails are very strong and healthy, and they can grow very long, but as soon as I use nail polish they become brittle and start peeling, and they continue to do so until all the "old nail" has grown out and the whole nail is replaced. I have tried 5-free, 7-free and water based polish, and I have also tried different removers. But no matter what, my nails get ruined every time I use polish. Do you have any suggestions what I should do?

Here are my top suggestions for addressing nail plate surface peeling:

Never peel nail polish from your nail plate. This will weaken the surface layers. These weakened surface layers will eventually peel away when they reach the free edge. It may be months later, so the peeler forgot what they did. If you peel nail polish from your nails in May, expect the cell layers on the top-side of the free edge will peel in until September when those damaged nail cells finally grow off the free edge. Doing this repeatedly can lead to a state of constant surface peeling. Some nail coatings adhere better, the longer you wear them. Long wearing nail polish has greater adhesion, so these are more likely to harm the surface when forcibly peeled off. Gently remove nail coatings, without the use of any force, e.g. no peeling, scraping, biting, etc. There is no such thing as gentle peeling or scraping, so don't fool yourself into thinking you "lightly scrape" or "peel carefully".

Keep your hands out of water. Each time you saturate the nail plate, surface layers swell apart and separate. Repeated soaking and drying cycles can weaken the bond between the upper layer and lower layers. Don't wash your hands too often. Yes, you can wash your hands too much. More than ten times a day can be hard on nails and skin. Soaps, cleansers and detergents can eventually strip away substances that help hold the surface cells to the underlying cells and this can lead to peeling. The same can occur when nails are exposed to cleaning solvents.

It doesn't matter if the polish is 5-free or 99-free, that's just "marketing" and doesn't say how good or how safe a nail polish is.

Don't think this is the problem or the solution. Buy high quality products, less expensive nail polishes are less expensive for a reason and often use inferior ingredients or are poorly formulated.

Keep nails shorter. The longer the nail plate, the more flexible the free edge, the more likely the polish will peel. This is especially true for those with thin, flexible nail plates. Cap the free edge by wrapping the base, color and topcoat around the free edge to the underside and give it some extra protection. Avoid skin contact. Wear gloves when digging or working with hands. Treat your nails like jewels, and don't use them as tools.

Protect nails from the sun. The nail plate has a high natural SPF, so the nail bed is protected from UV exposure, but that means the upper layers absorb the UV. Long periods of excessive sunlight can weaken surface layers and cause them to come apart.

Don't over file or buff the nail plate. Too much filing or buffing thins the nail and it is much harder for polish to adhere to thin nails, than thicker nails. Don't try to file away so-called "ridges". The nail plate can't grow ridges, those are actually grooves. Therefore, filing the plate smooth reduces and thins the entire nail plate to the match the thinness of the deepest groove. That's trouble waiting to happen- so don't do it.

Use nail oils, they can help reduce surface brittleness and help toughen the bonds between the surface and lower layers of nail cells, however make sure to remove surface oils before applying any nail coating. Nail oils also absorb into the plate to make it more resistant to excessive water absorption.

Avoid over exposure to solvents. Solvents can remove surface oils and may also leach out substances that help to cement nail cell layers together. Occasional use of solvent-containing polish removers won't have much effect on normal nails, but may have a noticeable effect on plates that have weakened adhesion due to other factors described above.

Peeling nail plates are never a sign of allergic reactions. The nail plate is not living and does NOT have an immune system, so allergic reactions are not possible.

Natural Nail Structure

52:1 A debate has come up about Beau's lines. I know that it typically occurs due to illness. But can trauma to the nail fold, injury or even a client habitually rubbing and picking at the eponychium area cause Beaus Lines? Or can matrix injury cause the nail plate to grow in the Beau's line-like waves?"

Beau's Lines are a "groove" or depression in the plate that extends across the width of the plate and not down the length. Such depressions in the plate are caused when the entire nail matrix is forced to slow down the production of new nail cells. While the slow-down continues, the new nail plate growth will continue to be thin. However, when new nail cell production returns to normal, the nail plate gradually returns to its normal thickness.

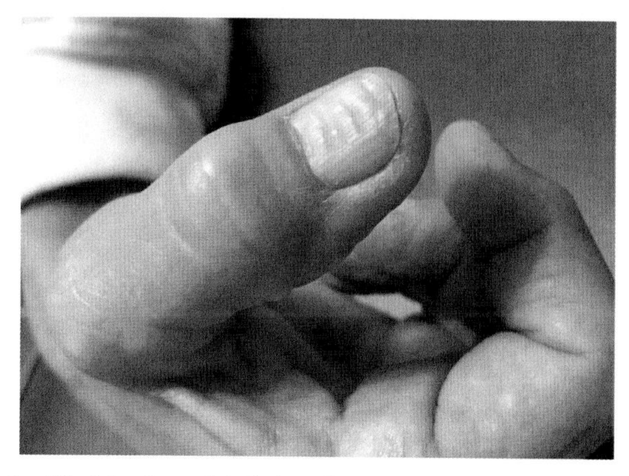

Image 6: Habit tic, which is often mistaken for Beaus Lines, which can have a similar appearance. (Photo credit Tina Karres Cornelius, North Carolina)

Therefore, Beaus lines are caused by internal health issues or medical conditions lasting a month or more- not by shorter term health issues, such as colds or flu. Serious accidents, surgical procedures, long-term malnutrition or heavy dieting, serious infections, heart attacks, uncontrolled diabetes or taking chemotherapy drugs can all cause internal health issues which can lead to the appearance of Beaus lines. The disappearance of Beau's lines indicates the body has healed itself or otherwise recovered from the original medical condition. The appearance of Beau's lines is also an indication that the nail plate will have a lower ability to resist infection. Bacterial or fungal nail infections are more likely when Beaus lines appear, because the nail plate may also lose much of its natural ability to resist infections.

Damage to the matrix can cause lengthwise splitting of the nail plate or loss of the entire nail plate, but it is not going to cause a Beau's line. Habitually rubbing or picking the nail plates can cause the shape of the nail plate to deform, but this is called a "Habit Tic" and it is not a cause of Beaus Lines. Beaus lines are from internal disorders while a habit tic is the result of repeated external damage. In any case, it is important to remember NEVER to attempt to file the nail plate smooth to remove Beaus Lines. This may cause significant harm to the existing nail plate and lead to over thinning and weakening the plate, which could make a nail infection more likely to occur.

56:1 What causes the nail plate to turn black or darkly colored?

Several things can cause the nail plate to turn black, or at least darkly colored. One of the most common causes of darkly colored nail plates is physical damage that leads to bruising of the nail bed. Hard bangs or knocks to the nail plate can lead to breaks in the tiny blood vessels in the nail bed, which can cause them to leak. These are called splinter hemorrhages and most often appear like tiny black lines running along with the direction of nail growth. When the damage is more severe, larger amounts of blood collect underneath the plate. A significant amount of damage could lead to blackening of the entire nail plate. The blood on the nail bed may appear to be red at first, but will eventually turn black. This black stain could persist for months after the damaged nail bed heals.

Certain drugs can affect the matrix cells and this can lead to dark discolorations of the nail plate. For instance, drugs containing metallic silver can cause dark black stains, as well as a surgical scrub used in hospitals which contain the ingredient "chlorohexine". Diseases of the nail or illness somewhere else in the body may also cause the nail matrix cells to produce darkly discolored nail plates. For example, diabetes can cause the nail plate to yellow and nail psoriasis often leads to nail plates that are pitted, more porous and pick up stains more easily. Chronic diseases such as kidney or liver problems can cause nails to blacken. Tumors in the nail matrix or surrounding areas can also cause dark discolorations in the nail plate, which is something that nail technicians should watch for. However, it is fairly common and normal for people with very dark skin to develop black strips that run the length of the nail plate. This can happen because excessive amounts of melanin in the skin can find its way into the nail plate. When this occurs on nails of those with African or Asian ancestries, this may be normal. However, when this occurs on nail plates of those with light colored skin such as Caucasians, in rare instances this can be a sign of potential trouble. Such nail plates should be examined by a medical professional to determine the causes, which could be one of many disorders. Finally, this change in coloration could indicate a nail infection, so in any case this type of change should be cause for concern. I recommend refraining from treating this nail until it can be examined by a medical professional to determine if an infection is present.

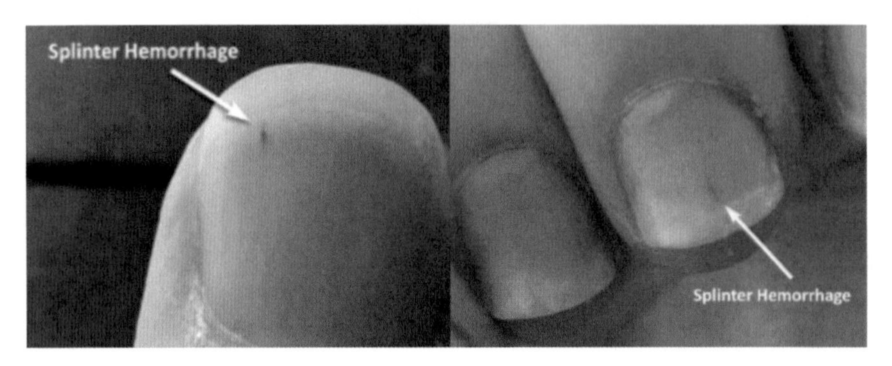

Image 7 & 8: Splinter hemorrhages are most often caused by physical trauma to the nail bed causing bleed underneath the plate. (Photo credits Becky Muffick, Dorango, CO (left) and Caroline Evans, East Sussex, England (right).)

56:2 What would cause a nail plate to turn grey?

Certain medications can do this, but not many. Also, the mosquito-borne disease called malaria, is also known for creating gray nail plates. Over use of certain nail hardeners may cause a gray appearance, particularly those that contain methylene glycol, which is also sometimes called formaldehyde. A more likely reason for grayish colorations could be related to the oxygen levels in the nail bed. When the blood in the nail bed is low on oxygen, the blood appears to be blue. If the nail plate already has a yellowish cast, the combination of blue and yellow can create a grayish-blue discoloration. So, gray nails could indicate something as simple as poor circulation to the fingers, but if the color darkens, this could suggest a more serious problem that should be investigated by a qualified medical professional.

63: 2 "I am not clear about how to work on nails with Pterygium. Can it be pushed back gently? I have several clients with skin that extends well onto the nail plate.

Many don't understand what pterygium is, but this nail professional apparently does. "Pterygium" is any abnormal growth of skin that becomes stretched. This can occur on any part of the body, on the eye, fingers, toes, elbows, etc. When it occurs on the nail plate, it is considered an abnormal growth of skin. A true case of nail pterygium is most often a result of disease or injury to the eponychium, as shown in the image below and occurs in three stages.

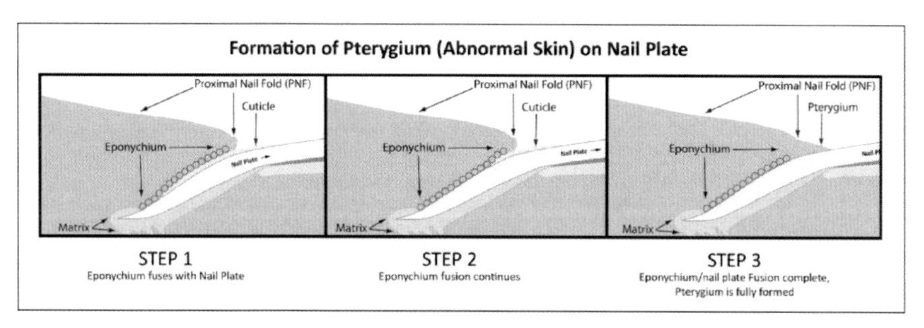

Image 9: The eponychium slowly fuses to the nail plate, in three stages to create nail pterygium and prevent cuticle from forming.

Researchers who study these problems believe some types of pterygium occur because the proximal nail fold fuses with the nail plate, as shown. The fusion is progressive until eventually all of the eponychium is fused to the nail plate. From this point, the proximal nail fold has become fused to the nail plate and will be stretched toward the free edge along with the nail plate in a triangular shape. Nail professionals should not try to cut, abrade, remove or reduce the pterygium, since that would be a medical treatment and outside the scope of allowed practices.

The fusion is thought to occur due to injury or disease, e.g. burns, physical trauma, lichen planus, and certain medical conditions also cause this abnormal growth. Such conditions should be referred to a doctor for examination if they have an unhealthy appearance. Pterygium should not be cut away by nail technicians, since it can bleed and become infected. It can be softened and conditioned, e.g. hot oil manicures.

Hardening and thickening of the proximal nail fold is not pterygium. This tissue hardening, also referred to by some nail technicians as “overgrowth”, is often created by the nail service, e.g. by cutting, abrasion, or rough treatment. This is much like callous formation which occurs when the skin is repeatedly injured. If that is the case, the nail professional should eliminate the parts of the service that are causing the skin damage. Then, eventually this condition may resolve itself. The client may be able to have gentle manicures, without disrupting the pterygium, and eventually the tissue may recover from previous harsh treatment that caused its formation in the first place.

This recovery can typically take a month to several months to occur. If the condition worsens, e.g. becomes red, swollen, tender or shows other signs of inflammation the client should be immediately referred to a medical practitioner; preferably a dermatologist (best for hands) or podiatrist (best for toes), so they can get a proper evaluation and treatment if needed. No nail technician should take it upon themselves to "diagnose", "treat" or prescribe treatment for this or any other medical condition. Practicing medicine without a medical license is forbidden by most countries, if not all. To learn more about pterygium and how it forms, See Volume ll, Topic 2: Proximal nail fold damage and pterygium formation.

69:5 Is the eponychium part of the proximal nail fold or is the proximal nail fold part of the eponychium. I'm confused.

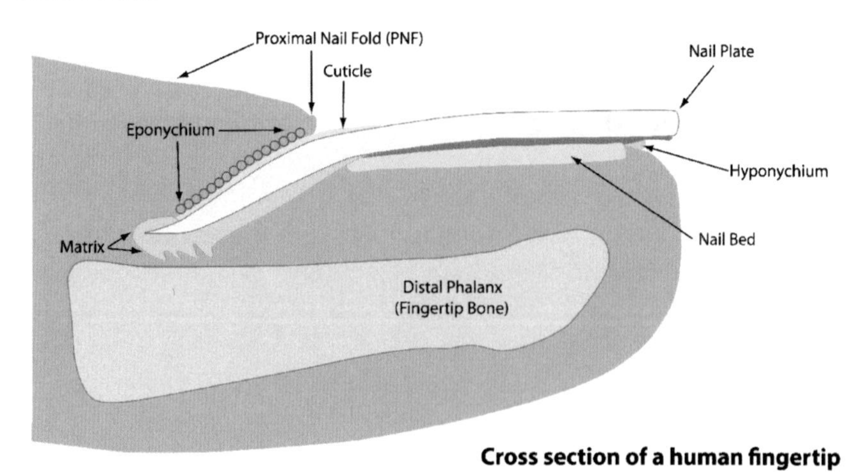

Image 10

You are not alone. Nail anatomy terminology is confusing. In fact, everyone seems to be confused about the names for the parts of the natural nail, even doctors and scientists aren't sure which terms to use. It can be very difficult to know the facts, but the facts are what we need or the confusion will continue and we'll never have any clarity or agreement. My goal is to come to some type of agreement on what we call the parts of the nail. To reach this goal, I've been working with many top nail educators to address this huge problem that plagues the industry. To get the ball rolling, I created a drawing of a nail as a cross-section image and labeled it based on a strict interpretation of the medical definitions. I shared that first version with many of you, as well as top scientists, dermatologists, podiatrists and pathologists and asked them to poke holes in the diagram. I expected to hear a wide range of different opinions, but I was looking for hard facts and would accept nothing less.

As a result, the esteemed Dr. Robert Baran, a well-known dermatology expert on nails, was instrumental in directing me to the conclusive evidence that finally settled a long-standing debate about the eponychium, cuticle and proximal nail fold. Here are the up-to-the-minute facts:

The "eponychium" is defined in medical literature as the skin that covers the nail matrix and is responsible for development of the cuticle tissue that adheres to the top of the nail plate. The proximal nail fold is defined as the fold of skin at the base of the nail plate. That much is clear, but where does the eponychium end and the proximal nail fold begin? Researchers have answered this question definitively by isolating and identifying the cuticle forming area.

It was discovered that the eponychium is a much thinner layer than suspected- in fact it is surprisingly thin- approximate 0.1-0.15 mm or about 0.004-006 inches thick! Rather amazingly, all nail cuticle tissue comes from this thin layer of cells. How can such a surprisingly thin area make all that cuticle tissue? The prevailing thought is that the eponychium must be made of a specialized type of cell called an "adult stem cell". The same type of stem cells are suspected to form the nail matrix and produce nail plate cells. Research is underway to verify that both the eponychium and nail matrix are composed of adult stem cells. These stem cells behave like factories to produce cells that make up the nail plate and cuticle. Dr. Baran's insightful description is that the proximal nail fold should be viewed as a "flap" of skin that covers the matrix area and its underside is thinly covered by the eponychium. I think that paints a good picture. So, in short, the eponychium creates and releases the cuticle as a thin layer of dead tissue that will ride the nail plate and emerge from underneath the proximal nail fold. Its function is to form a protective seal that will prevent pathogens from entering to cause an infection of the matrix area. The cuticle tissue should NOT be confused with the eponychium or the proximal nail fold. They are each very different.

73:2 My daughter is two and a half years old and has been sucking her thumb since birth. As you know children's nails grow incredibly fast, but this one never ever grows and never needs filing. Is this due to sucking the moisture from the nail?

I would need a lot more information before I could understand why the nail plate wasn't growing on this thumb, but even so, it would not be possible to suck the moisture from the nail by mouth. The nail plate is not like a living plant that needs moisture for growth. The amounts of moisture in the nail plate is not responsible for nail growth. In fact, the opposite would occur, the nail would have a lot of moisture since this would be transferring from the mouth into the nail plate and absorbed. The mouth is very moist. Besides, the nail matrix controls the growth rate. This helps explain why cutting the nail plate does NOT make the nails grow faster or slower.

73: 3 Why can thumb sucking make the nail plate grow in distorted shapes?

The nail plate may seem solid and fixed in shape, however, recall that some solids can flow and change shape. A glacier of ice is an example of a flowable, shape-changing solid. The nail plate is also a flowable solid. Prolonged or repeated pressure can cause the nail plate to flow and slowly change shape over time. Thumb sucking can put enough pressure on the plate to deform it, but not in a day or even a week. However, over many weeks or months, when enough constant pressure is applied, the nail plate will deform and adjust to the pressure by permanently changing shape. For instance, those who constantly flick their nail plate with another finger can distort the shape of the nail plate. Such conditions are called "Habit Tic". This shape shifting of the nail plate will relieve some of the external pressure applied to the plate. In this case, the applied force comes from the child's mouth, but the same can occur when shoes are worn too tight, especially when toes are squeezed into small areas, such as pointed toes.

74:4 I know the hyponychium and onychodermal band form a protective seal for the nail bed, but I wondered why there are two types of seals? How are these different from the solehorn?

I know this is confusing because many of the current illustrations are unclear and clarity is what is needed. Sandwiched between the nail plate and nail bed is an important, but very thin layer of tissue called the nail bed epithelium.

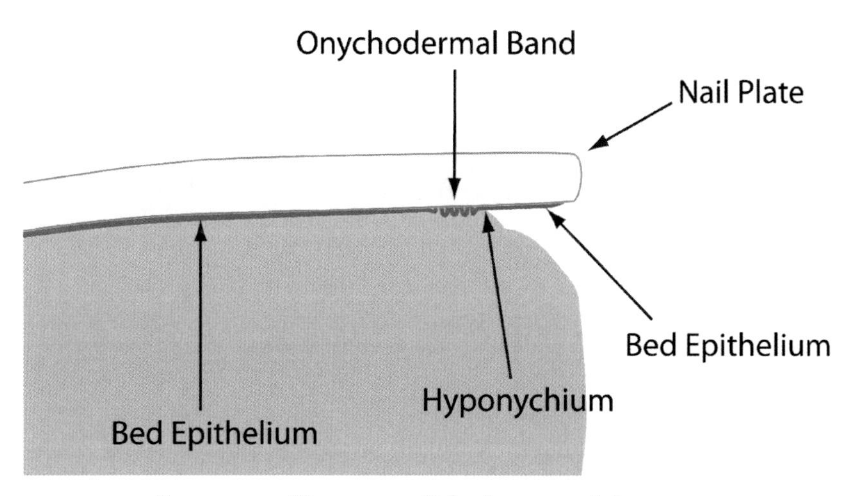

Image 11: Close-up of the hyponychium area.

Many don't understand, but this paper-thin layer of tissue helps to guide the nail plate as it grows. Interestingly, the bed epithelium adheres only to the underside of the nail plate and not to the nail bed itself. This allows the plate to glide smoothly across the nail bed allowing it to eventually reach the free edge and beyond. As the nail plate moves, the bed epithelium continues to tightly adhere to the underside of the nail plate, even as it moves past the finger and can be seen on the underside of the nail, still firmly attached. This tissue is usually removed during a manicure when cleaning up under the free edge. The tightly adhered tissue on the underside of the nail plate is still the "bed epithelium", but some feel a need to give it a new name and they call it "solehorn". I don't agree that this old-fashioned name should be used. Why? This term existed before the true source of this tissue was properly understood. The solehorn does not simply emerge out of nowhere nor does it come from the hyponychium, so why should it be given a different name? This is still bed epithelium tissue and should be called by its proper name.

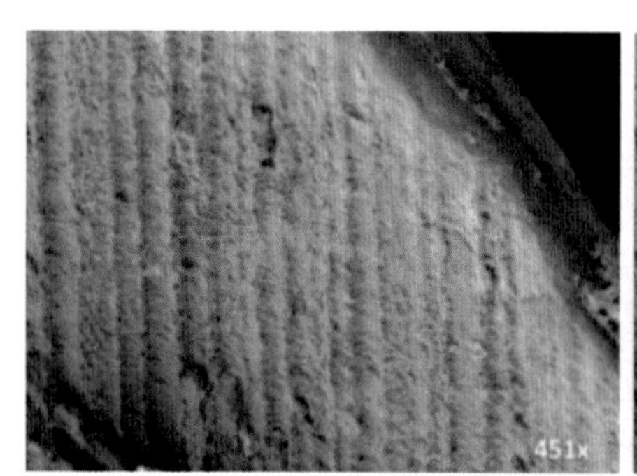

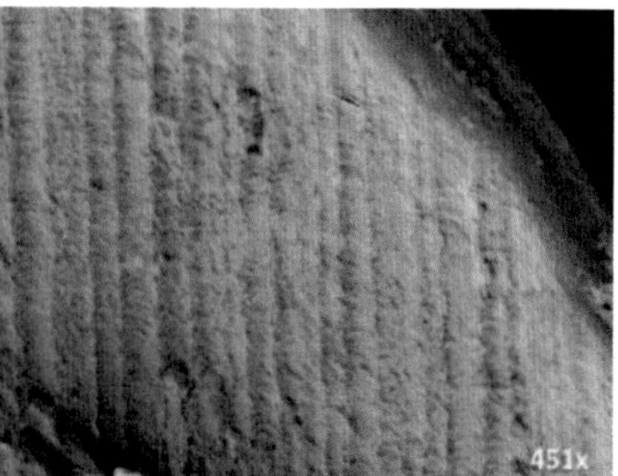

*Image 12 A: Bed epithelium clinging to the underside of the free edge. Image 12 B: 3D image of of the bed epithelium, **Red/Cyan** 3D glasses required*

The onychodermal band is NOT another kind of tissue even though it does form a distinct region. The onychodermal band is also created by the bed epithelium. How? As the bed epithelium moves with the nail plate, it must somehow squeeze past the hyponychium, which is the seal under the free edge that prevents pathogens from infecting the nail bed. It's a tight squeeze getting past this seal, but the bed epithelium manages to squeeze past. In doing so, it becomes bunched up and doesn't flow smoothly past the hyponychium. This "bunching" causes a rippling effect which creates a barrier that prevents penetration of substances from seeping past the hyponychium, so this helps to protect the nail bed. In the zone preceding the hyponychium is where this bunch up of the bed epithelium occurs. This bunched up, rippled tissue produces a greyish zone called the onychodermal band. Therefore, this is only a visual effect caused by the bunching up of the bed epithelium as it grows past the hyponychium. You can't see the onychodermal band by looking under the free edge, it's not there. It is located underneath the nail plate and on the nail bed and can only be seen by looking though the nail plate and observing the grayish band that occurs just before the hyponychium.

Natural Nail Service/Treatment

52:4 Can you give me your opinion on using an electric file with a 180-grit mandrel band on the natural nail please? I have always been led to understand that it is not good practice to use an electric file on natural nails but recently it has been demonstrated to me and it looks no different than a hand file. I teach at college and privately, can you give me an answer on what is acceptable.

I understand that electric files do have some value, but don't see the advantage to using an e-file with a 180-grit sanding band to prepare the natural nail plate. It can be done if one is always extremely careful, but that doesn't always happen. One slip of the hand or a break in concentration and a client's nails could be seriously damaged. I could cut my lawn with a chain saw, but what is the advantage? And just because I can, doesn't mean I should! Using an electric file in this fashion is likely to cause over-filing. Over-filing is a major contributor to much of the nail damage clients suffer.

Using such techniques is common in salons that do "factory nails". Speed seems to be the main advantage, but that should not be more important than safety. When a nail tech uses an electric-file and a 180 grit sanding band directly on the natural nail plate, in my opinion, it's for the nail technician's convenience and NOT for the client's benefit. It is easy for this educator to say the nail plate looks the same as after hand filing, but I suspect she didn't measure the thickness of the natural nail before and after use. Many nail technicians tend to over file the natural nail and they don't need another method that makes over filing even easier to do. I don't condone anything that promotes more over filling. Very little of the nail plate needs to be removed from the nail plate and I can't imagine a power tool is needed directly on the nail plate. Those who promote this practice aren't properly considering the client's nails, or they would not use such a powerful and potentially aggressive tool on the nail plate. In my view, when it comes to filing on the natural nail, less is best!

57:1 If I use pure 99% isopropyl alcohol to cleanse the nail plate, can I say for certain that it will remove all the surface oils and moisture from the nail plate?

You are right to call them surface oils. The nail plate contains a lot of oil, but we only want to remove surface oils because they can block adhesion. We are making a simplification when we say the nail plate contains "oil". Nail oil contains many different oily substances, and some are much more difficult to remove from the nail plate than others. The natural oils in the nail plate are a complex blend of more than a dozen different waxes and oily substances. Most of these substances are secreted by the tissue surrounding the nail plate, but some are transferred from the hair to the nails when they come into contact with each other. Not all oily substances found on the nail plate may be removed when the nail plate is cleaned with either acetone or alcohol or ethyl acetate, the most common solvents used.

That's because not all types of nail plate oily substances are soluble in every solvent. Some are more soluble in acetone, while others are more soluble in isopropyl alcohol. In other words, different solvents selectively remove various contaminants from the surface of the nail plate. Some solvents can leave behind contaminants that may fill up and/or block the various tiny spaces where the nail coating would normally seep into, harden and anchor itself more firmly to the surface of the nail plate. This helps explain why cleansing the nail plate is so important to good adhesion. A scientifically designed nail surface cleanser will contain a synergistic blend of solvents and other beneficial additive ingredients that are more likely to do a superior job when compared to using isopropyl alcohol alone. That's why I recommend using professionally designed nail cleaners and avoiding less effective substitutes.

68:6 "Should e-files be used to remove hardened tissue at the base of the nail plate or is this an overly aggressive technique".

Invasive manicures use high speed power tools to abrade away living skin, simply to save time and effort for the nail technicians. These risky procedures are NOT in the client's best interest, because this opens the skin like a window to allow microbes and infectious pathogens to infect the living skin. I recommend "non-invasive" manicures which do not abrade the living skin. Clients don't realize that the hardened tissue on the proximal nail fold is there only because their nail tech uses overly aggressive techniques that damage living skin to create the hardened tissue. Clients are not happy when they learn their nail tech is the cause of this hardened tissue. They get upset when I explain their skin would be much healthier and they would be safer if they avoid these risky procedures. I explain to them that their nails would be in better condition if the nail tech was better trained and more responsible.

When they learn this, they usually seek out a better trained and more knowledgeable nail professional who doesn't invade the skin. Unfortunately for those who use invasive techniques, it is my aim to better educate the public so they will avoid invasive manicures.

68: 7 Can you explain the downside risks of filing the side walls of the nail plate?

When the side walls are filed, two important problems can occur. When the nail plate width is reduced, the overall strength of the nail plate is also reduced and weakened. Not only that, the plate becomes overly flexible and will bend too easily. Bending back and forth will weaken the nail plate and can cause stress fractures to form near the sidewalls where the plate is flexing. Of course, this can lead to other problems such as a loss of adhesion of the nail coating and/or breaking nail plates. This can cause the shape of the nail plate to curl more easily, either upward into a ski jump or ski hop nails, as they are often called, or they can bend sharply downward as they grow past the free edge. In these cases, the integrity of the nail plate has been compromised. An even greater issue is the increased potential for allergic skin irritation or allergies of the side wall tissues, since abraded skin is easier for substances to penetrate. When this soft tissue is abraded by the file, it can cause weakening of the skin defenses. Exposure of this abraded tissue to nail coating products is of great concern, since prolonged or repeated exposure to this damaged area increases the risk of adverse skin reactions. Clients pay nail professionals to beautify their nails, but not at the expense of the nail or skin health. Pay close attention to this issue and be sure you aren't overfilling the sidewalls or the top surface of the nail plate. Over filing is one of the biggest challenges facing the nail industry and a major reason why many potential clients avoid nail salon services. When it comes to filing "Less is Best"

71:1 If lipids do not dissolve in water - do they dissolve in alcohol? Would cleaning a nail plate with alcohol significantly reduce the amount of lipids?

Lipids are another name for oily substances called Fatty Acids and their derivatives. They occur in most natural vegetable oils. Lipids are oily substances and are not soluble in water. For this discussion, it is important to know there are many types of alcohols, more than several dozen different types are used in cosmetics. The simple, low molecular weight alcohols, like ethyl or isopropyl alcohol, can dissolve and remove most lipids from the surface of the nail plate. However, they are more easily dissolved by other types of solvents, namely acetone or ethyl acetate, which is why these solvents are often used in nail cleaning or preparation products. Of all these, I believe that acetone is a superior solvent for lipids found in the surface of the nail plate. Using any of these solvents with a fiber-free pad is a great way to remove surfaces oils and lipids. But a simple quick "swipe" is not nearly enough. To remove surface oils and improve adhesion, nail technicians should use a solvent wetted cotton pad and scrub the nail plate as if removing red nail polish. This creates the necessary mechanical action needed to pull these surface contaminants from the tiny nooks and crannies on the nail plate. In my view, nail cleansers created by manufacturers are likely to do a better job, since they often contain synergistically blended solvents and other additives to improve cleaning.

74:2 I am confused. I see the term "moisturize" used when oils are applied to skin and nails. I don't see how oil can moisturize. Doesn't the term mean to add water, not oil and how are they different from humectants?

I agree, these terms are confusing. The original definition of moisturize is to reverse dryness, which is caused by the lack of water- aka moisture. Some consider the absorption of any liquid into a solid to be moisturizing. And many marketers misuse this term to mean making skin more soft and supple, which is actually a result of adding moisture to skin. So, the meaning of this word has become confused by cosmetic marketing. Many are taught that oils are moisturizing, which seems contradictory because natural oils contain no water. But I do agree that oils on the skin and nails can increase the water content of skin and nails. This occurs because the oils can seal the surface and prevent moisture from escaping. Therefore, the moisture builds up underneath and the water content of the nail plate will increase if nail oils are applied. However, when nail polish, UV gels or any other type of nail coating are applied, the same occurs.

The water content of the natural nails can be doubled by wearing nail coatings. When this occurs, the increased moisture will make the nail plate much more flexible, supple and soft. Even so, no one would claim that nail polish is moisturizing. When you hear this word, realize it can mean many things and it can mean nothing at all, so don't be too impressed by claims of moisturizing. Rinsing your hands and nails in water is moisturizing as well. However, there are times when this term is more meaningful. Some ingredients can absorb into nails, hair or skin and act as moisture magnets. These ingredients bind with water and will hold on to it tightly. These types of ingredients not only increase the moisture content, they can prevent the water from escaping or evaporating. Sodium hyaluronate is a perfect example. It is added to cosmetics because this substance is distributed widely in tissues, including the eyeball. It binds moisture and acts like an internal tissue lubricant. Humectants can draw excess moisture from the air or surrounding tissues, which is why they are called moisture magnets. They work to increase the moisture content of their surroundings. Other common examples are panthenol, glycerin, urea and propylene or butylene glycol. The opposite of a humectant is a desiccant. Desiccants are a different type of moisture magnet, ones that have a drying effect. They absorb moisture from their surroundings and decrease the moisture content. Examples of desiccants are table salt, silica gel, bentonite clay and calcium chloride.

Pedicure and Foot Related

56:3 I was taught not to give a pedicure to a client with athletes' foot. Rather suggest that they see a physician but often they become very defensive about it. How can I help without offending the client?

Yes, I see your problem. Athlete's foot is perceived as minor in the client's eyes, and I must agree, it is minor, but it is also contagious. Try to explain to the client that they must have become infected by someone else because these organisms are easy to spread. Ask them to think about the inconvenience and discomfort this infection has caused. These organisms are spread by people who could have taken steps to prevent the transmission to others, but they didn't. Now countless others will become infected and suffer in the same fashion. Ask the client how they feel about the careless or clueless person who transmitted the infection? Then explain that you don't want to risk transmitting this to others.

Salons should NOT be part of the chain of transmission of these infections. That is counter to everything that salons are about. Besides, this is absolutely terrible for business. Salons are supposed to be a safe haven for clients, right? They can't be a safe haven for both those with foot infections and those with healthy feet. That's why those with unhealthy feet should see a medical professional, such as a podiatrist or dermatologist. Those with healthy feet should see a nail technician or pedicurist. You can ask her to consult with a local pharmacist for a treatment, since this type of fungal infection is fairly easy to cure.

64: 1 Could a client develop athletes foot or other fungal infections symptoms within a few days after getting a pedicure?

According to research I've seen on this subject, there is delayed period between infection and development of symptoms. This is true for all transmittable disease, including fungal infections of the feet. Athletes' foot is a fungal infection caused by an organism called tinea pedis. The delay before appearance of visible signs of infection is called the incubation period. During the incubation period, the infection is actively spreading, but not yet producing noticeable symptoms. Noticeable symptoms won't appear until these organisms have multiplied to high enough concentrations to damage the skin. At that point, a healthy immune system will spring into action.

The immune system releases substances called histamines into the blood which expand the blood vessels so they can deliver more blood directly where it is needed most. A side effect of histamines is localized skin itching and redness. How long before the first symptoms appear? This can depend on several factors, but typically, athlete's foot has an incubation period of about one week and symptoms typically take about 7 days to appear, but that's only if the skin was previously healthy and in good condition. If the skin were already damaged or compromised, infection could occur a few days' sooner, possibly 4 or 5 days. Symptoms are unlikely to occur any sooner than this.

Ingredients/Products

52:2 What does "cosmetic grade" mean? I have formulated a nail art paint with a non-toxic airbrush medium that claims that there are no health hazards for intended use, but warns to avoid ingestion, excessive skin contact, and inhalation of spraying mists, sanding dusts, and concentrated vapors. The company claims the formula is a "proprietary blend" and they won't tell me the ingredients. I am mixing the medium with pigments that are FDA approved for use in cosmetics. What all must I do to ensure this blend is safe and can be sold for this use?

This question covers a lot of territory, so I couldn't tell you all that you need to know, but I'll do my best to answer. I'm sure others have similar questions about this. First, "non-toxic" is a marketing term with no real definition, so don't be impressed by this claim. Anyone who sells a cosmetic product is required to know the ingredients, so they can list them on the label and on the product's Safety Data Sheet or SDS (aka MSDS). They also must make sure the ingredients are not prohibited and determine the proper ingredient name to place on the label. This information is obtained from the International Dictionary of Cosmetic Ingredients (INCI Dictionary) it's a list of names for cosmetic ingredients. Product manufacturers must ensure that all ingredients used are listed on the label and the correct INCI name is used for all ingredients. It is unlikely that any cosmetic product would be allowed to avoid listing the ingredients by claiming that it's a proprietary blend unless you have trade secret status and that is a hard claim to make if you don't know the ingredients. Even so, without the ingredient listing, the product couldn't be sold outside the US, because Europe, Canada, Australia, Japan will all insist on knowing the ingredients. Also, anyone who brings a new cosmetic product to market is expected to ensure all ingredients are safe for cosmetic use and not restricted from use in cosmetics.

Since the nail technician who asked this question is in the USA, it would be very important for her to ensure the colorants and pigments are FDA approved for use, specifically in cosmetics. That's what is meant when the term "cosmetic grade" is used. To keep from being repetitive, I'll use the word "colorants" to mean anything that is intended to impart color, including liquid dyes and solid pigments. It is very important to understand that the FDA and other agencies in different countries have approved a large number of colorants, but most are not approved for cosmetic uses. They represent a different "grade of material" that is considered non-cosmetic grade.

This is usually due to the chemical composition of the substance and how it is intended to be used. Non-cosmetic colorants may not be safe for use in nail products, so they should be avoided. For instance, if a non-cosmetic pigment which is based on nickel were used in nail products, it could trigger allergic reactions in those with nickel allergies. It's estimated that 1 in 10 women in the US have nickel allergies, largely from wearing inexpensive jewelry that is plated with nickel... so you see how this could be a problem. Only a relatively few colorants are considered safe for use in cosmetics and many countries closely regulate and control the use of all cosmetic colorants. Only those specifically approved for use in cosmetics may be added to any cosmetic product, including artificial nail products. Those that are not specifically approved for cosmetics are NOT allowed to be added to cosmetics. The EU and other countries also have requirements similar to the US, so this is a universally accepted concept.

Product sellers are also required to develop and actively distribute product Safety Data Sheets to nail professionals who use their products and to update these sheets regularly. Those are the basic requirements, but it is also important to know that all sellers are responsible to provide safe products, as well as to provide safe usage directions, and any warnings or precautions. Warnings and precautions MUST be prominently displayed, which means created so that nail technicians can see and read them. Everyone in the USA who sells cosmetics must package them in accordance with the Fair Packaging and Labeling Act, which is a federal labeling act. Other countries have their own similar packaging standards. More information on the this act can be found at the US Federal Trade Commission's website and you use the handy search box feature search for the "Fair Packaging and Labeling Act".

BTW: The video series "Face-to-Face with Doug Schoon" website also has a handy search box feature that allows you to search for key terms and find Episodes that talk about a particular topic. For instance, use the term "glitter" and you'll find episodes where I have talked about it.

My recommendation to anyone thinking of selling a cosmetic product into the nail industry is to find a good regulatory consultant to ensure that your products and packaging meet the requirements of each place/region where your products are sold. That is very important. You can check with one of the many cosmetic associations such as the ICMAD or PCPC, for a referral. ICMAD is an acronym for Independent Cosmetic Manufacturers and Distributors. The PCPC is the Personal Care Products Council. Both have websites and both can refer you to knowledgeable regulatory specialists. There are rules and regulations in every country which must be adhered to. Now you can see it's a myth that cosmetics aren't regulated. Ha! There are so many regulations, some think too many, but for some there are never enough regulations. Some are constantly trying to force unnecessary regulations on top of the existing ones, which in my view is a foolish waste of time and resources.

Cosmetics are more heavily regulated than some activist's groups would like you to believe. I'm sure some of you are saying, I had no idea it was so complicated. And I'll tell you the process is far more complicated that it seems. That's why I recommend that anyone selling cosmetic products seek the help of a professional who specializes in cosmetic regulations and make sure you are doing things right. Regulations and standards for arts and crafts products are entirely different from cosmetic requirements. Some confuse the two and this should be avoided.

Here's what I mean- some have asked me why it isn't safe to use art and craft colorants or glitters when the product's website and literature and Safety Data Sheet say they are "safe" for use. Some don't understand that this means the colorants are safe for the "intended use" arts and crafts and that does NOT include cosmetic use. Just because these are safe for use in a craft project, doesn't mean they are safe for nail products or other cosmetics. Inhaling filings and dusts that containing non-cosmetic colorants could cause problems for nail technicians. And overexposure can lead to rather serious allergic reactions. Whenever an ingredient is purchased for use in a cosmetic product, the manufacturer should clarify to all companies that the ingredient is for use in a cosmetic and explain the intended purpose so the ingredient manufacturer could let them know if it is approved for that usage. Anyone who manufactures or sells cosmetics are required to do many things, most of which help to ensure the products are safe when used as directed. Manufacturing and/or selling a cosmetic is serious business with serious responsibilities, which is why I recommend that all involved ensure they are compliant with all requirements.

53:2 I try to do my research on the Internet to find out how to use different products, but I get so much conflicting information that I don't know what is true or correct. What can I do?

I'm also very concerned that the Internet has become a place where nail technicians and hobbyists regularly share misinformation and teach each other their bad habits. The Internet allows incorrect information to be shared on a grand scale. Much of this incorrect information comes from veteran nail professionals who learned their information from unreliable sources when they were in nail school. Now they teach what they learned to others in YouTube videos and blogs.

I'm especially concerned because the incorrect information VASTLY out numbers the correct information on the Internet research, nail technicians are FAR more likely to be deceived or confused and NOT to become better educated. This means on-line nail technicians are more likely to learn incorrect information and poor techniques. I recently saw a video from a well-known nail educator who discussed the virtues of performing cuticle care after applying artificial nails, rather than before. Her theory was that if done before, this could lead to lifting. Clearly this veteran doesn't know that the cuticle is dead tissue on the nail plate, which must be removed BEFORE applying artificial nail coatings, it's no wonder so many are confused.

In another video, a different nail tech tells viewers that MMA can kill clients... which is clearly ludicrous. MMA is the most popular bone cement in the world and has been used to mend seriously broken bones and hips for more than 25 years. It's not a good for nails and shouldn't be used, but it doesn't kill people either. With so much misinformation available to nail techs today, it's not surprising that in general, the level of education for new nail professionals is lower now than I've ever seen during my 25 plus years of educating in this industry.

Largely because of the Internet and the ease with which misinformation can be spread, I am extremely concerned that if this continues, the nail industry as we know it will not survive. This reminds me of an old phrase, "Death by a Thousand Cuts". That is what I see happening to the nail industry, we suffer cut after cut and not much is done to stop this from happening. Activists unfairly attacking with their misinformation, the media provides misinformation to consumers to frighten them needlessly and many nail technicians don't know the facts. We would be wise to change this and we better do something quickly. The industry ignores each little cut, but the many cuts we've already suffered are accumulating rapidly. Nail technicians need to get more serious about learning and veterans must stop teaching out-of-date technique/opinions. One thing that would help is if more manufacturers stepped up to the plate to provide fact-based information and would refuse to market products by using fear. Everyone can help by sharing this information and encouraging the many thousands of under educated nail technicians to learn from my books and Internet video series. Death by a Thousand Cuts... its coming to the nail industry unless we get more serious and do something about it and soon.

54:1 Can we trust nail products that come out of China and why?

Of course some great products are made in China and all products exported into other countries must adhere to all the regulations of each country into which they are being imported. When products arrive at the shores of the European Union, Canada, the US or Australia, all cosmetic products have importation paper work and required certifications that must be checked before the products can enter.

If that inspection raises any concerns, the products may be quarantined until they can be checked in more detail and in some cases, they are tested to ensure they comply with any existing regulations. In general, these processes are effective and ensure that unsafe or illegal products are stopped at the borders and those that do not comply are returned or destroyed, depending on the reasons for non-compliance. However, there are cases where this doesn't occur, and I'd like to discuss these so that you are better informed and aware of a few potential issues. It is possible that something is allowed into the country and is sold, when it should have been rejected.

In general, products from China are safe, but there are those in China who attempt to get around the system and avoid detection by the authorities. One way this is done is by intentionally mislabeling products. For instance, the label does not list the actual ingredients that were used. The same happens with the Safety Data Sheets, which will also be incorrectly filled out with misleading information. This can be done to fool both the importing purchaser as well as the inspecting authorities. This helps explain why it is important to only purchase such products from reputable sources of distribution. They are responsible for scrutinizing all suppliers, including imported products to ensure they are getting what they paid for and they are responsible for ensuring the labels and SDS contain proper information. Purchasing from a well-known and reputable dealer is the best way to ensure this occurs, so I would recommend NOT buying nail products directly from China and trying to import them yourself. You may not get what you expect.

Another problem to be aware of are counterfeit cosmetic products, in fact this is a big problem. Counterfeit products are usually not the same product made and just sold at a lower cost, often counterfeiter's products are cleverly designed to mimic the appearance of a brand name product, but are made under unsanitary conditions, use inferior ingredients and may even contain unsafe substitutes. Some counterfeit products can be dangerous to use and most are sold over the Internet to unsuspecting buyer who incorrectly believed they are getting a "deal", when they are actually being duped out of their money. And these products don't just come from China, counterfeit products can be made anywhere. The best way to avoid these counterfeits is to only buy established, brand name products, and avoid those designed to be just-like the brand names- they usually are not and this is just a marketing ploy. Use the original products and avoid copy cats or clones.

Only buy these established brand name products from authorized distributors. Those are the ones authorized by the manufacturer to represent their product lines. Anyone not authorized to sell a brand is getting their products from questionable sources. How is it they can sell the products for less? Clearly something is not right and often these are low quality counterfeits or products that are nearing or have exceeded their recommended shelf-life. Don't be fooled. Many counterfeiters sell to unauthorized distributors who often unwittingly sell these fake products, usually over the

Internet. Authorized distributors are more likely to have the real products and unauthorized distributors selling at lower cost are more likely to be selling fake counterfeits. So, buyer beware. I recommend that you heed this advice and save your money and protect your health.

55:3 Is acetone a safe solvent for removing nail coatings? What should I use if the client is allergic to acetone?

Here are the facts, acetone occurs naturally in our bodies, in low concentrations, so it's not a foreign substance. Acetone is very unlikely to harm the body when used to remove nail coatings. But it can temporarily remove excessive amounts of surface oils which can cause the skin to appear "dried out". No one becomes allergic to acetone, despite what some say or believe. Some become allergic to uncured ingredients released by the acetone during removal, but improper curing of the nail coating is to blame, not the use of acetone. In fact, when a client develops a sensitivity related to acetone or other solvent removers, that is generally a strong indication that their nail coatings are not being properly cured. This can happen no matter what solvents are used and is a powerful reason for ensuring proper curing of nail coatings, not a reason for avoiding acetone. Therefore, acetone is a safe solvent for these applications, assuming it is used wisely, e.g. kept away from open flames, sparks, and not heated improperly, etc. Read the warnings and Safety Data Sheet and heed the directions for use, if you do, you should be safe.

66:1 Some nail products claim to help the nail plate grow stronger. Is this a valid claim?

Many are confused by "nail growth" claims. The facts are, some of these claims are real and some are illegal. Why? There is a really big difference between "making" something grow and "helping" something to grow. All nail professionals, should clearly understand the difference. For instance, if I put fertilizer on a plant, I "make" it grow faster. If I put a fence around it to keep the rabbits from eating the plant, I "help" it to grow taller. There is a really big difference here. The first one changes or alters the way the plant would normally grow, while the second works by protecting the plant, so it can grow naturally. It is illegal for cosmetics to claim that they MAKE the nail grow. Cosmetics are not allowed to change or alter the growth pattern or nails, hair or skin. Only drugs and foods can make these claims. Even so, cosmetics can "help" the nail plate to grow by making/keeping it in a stronger/tougher condition, which makes it more resistant to breakage. This is an important benefit that can allow the nail plate to survive daily rigors and to grow longer. Watch for the term "helps grow longer" and avoid those who illegally claim to make the nail "grow faster". No cosmetic can claim to accelerate nail growth. If they do, that would make their product a misbranded drug. Words are important. A single word can make all the difference between a real benefit and a phony claim.

67:2 Where does the term "FDA approved" come from and why do we see it in use?

The FDA is the US Food and Drug Administration and it approves medicines and medical equipment. It does NOT approve cosmetics or foods, but is of course concerned about their safety. No cosmetic or food or food supplement can claim to be FDA approved. This would be a false and misleading claim when it comes to cosmetics. The FDA does approve certain medical devices that are sometimes used in salons, e.g. medical lasers, but not for cosmetic products. Why do we see cosmetics claiming to be FDA approved? There is no shortage of tricksters and scammers trying to fool people into buying their products. They don't compete fairly and instead wrongly steal market share from competitors that do things properly and don't make inappropriate or false claims. In my view, any cosmetic company making FDA approved claims about their cosmetic products should be avoided because they are making false claims. They either are ignorant of the regulations or don't care. Either way, they are not a good source of information and probably can't be trusted to provide the facts.

68: 4 I heard an educator say in class that isopropyl alcohol is a carcinogenic, is this true?

No, that is false. Isopropyl alcohol is a very safe solvent and does not cause cancer. Neither does acetone, which is another common solvent. Both are safe for salon use. Unfortunately, when someone wants to scare you, but they don't have real facts, a common trick is to claim something causes cancer. Nail products don't cause cancer! There is a huge difference between "can" or "could" and "will". Deceivers pretend that "can" is the same thing as "will"- but it's a trick! A giant meteorite could crash into my house, but it is safe to say that is unlikely to happen, so why worry? There is no need to worry about carcinogens in nail products. That's not a realistic risk... and there is no evidence that this is a real threat.

69:2 I've read that Jojoba oil is able to penetrate the nail plate and that it helps to mix it with Vitamin E, to get it into the nail plate, as well. If this is a matter of molecular size, how is nail penetration possible with Vitamin E? Isn't that like trying to get a camel through a needle eye and wouldn't it just remain only on the surface?

First, I want to discuss some terminology. Vitamin E is a food supplement name for a substance that is properly named "Tocopherol". In other words, tocopherol is the correct cosmetic name that should appear on the label and the term vitamin should only be used for foods and ingested nutritional supplements. I'll use the correct cosmetic ingredient name, of course, and so should manufacturers and marketers. Tocopherol is a very large molecule and molecules of this size typically have a difficult time penetrating the nail plate, but the same could be said for jojoba oil. Certain molecules found in jojoba oil have a very long sleek shape, rather than a large, bulky shape, which can really make a big difference. I agree that vitamin E molecules would have a difficult time penetrating on their own. Most would likely sit on or near the surface of the nail plate, but that's not a bad thing. Why it that? This is where tocopherols protective action is most needed, near the surface where sunlight and oxygen exposure create unwanted and potentially destructive chemical reactions that lead to discoloration and embrittlement of the nail plate and/or nail coatings.

Tocopherol is most needed on or near the surface, and it is less useful deep inside the nail plate or enhancement. I have done testing with blends of tocopherol and various natural oils such as jojoba and avocado oil and these studies indicated tocopherol did indeed slowly penetrate the nail plate over time and it seemed to concentrate near the surface. These natural oils do act as carrier solvents for the tocopherol. That is NOT surprising since many vegetable oils naturally contain tocopherol. For example, wheat germ oil is the most concentrated natural source known for tocopherol. Therefore, I'm confident that some tocopherol does absorb, but not large amounts. However, large amounts of tocopherol aren't needed, since it is highly effective in very low concentrations. This helps explain why nail oils which contain tocopherol, must be used regularly to ensure that a protective amount of tocopherol is always present. Of course, tocopherol can also be blended into a monomer liquid and/or UV gel to provide similar protective effects throughout the coating. But, the real protective value of tocopherol is when it's concentrated at the upper surface. Therefore, using a high-quality nail oil with tocopherol is a valid approach to protecting the integrity of both natural and artificial nails.

Note: when tocopherol is added to artificial nail products, it's function is to protect the coating, not the nail plate. I am not aware of any evidence that demonstrates that any useful or significant amounts of "vitamins" can migrate from artificial nail coatings and absorb into the nail plate to provide benefits, so I recommend disregarding those claims until convincing evidence is provided.

69:4 What about cuticle guard products? I have not yet used them, but they look smart and could save time when cleaning after stamping. Are they safe?

Since these products are NOT designed to guard the cuticle, they are clearly misnamed. Many marketers don't understand or use proper terminology. This isn't all their fault, because even scientists and doctors are confused and the medical literature is contradictory. These products guard the proximal nail fold (PNF), not the "cuticle". Whatever you call them, I would not use one that contained latex. Many are allergic to this naturally occurring substance and should completely avoid them, as should anyone who has skin which is easily irritated or highly sensitive. Latex can cause permanent skin allergies, so be cautious. I recommend reviewing a product's Safety Data Sheet (SDS) to find out if any other risks are associated with a certain product and how to avoid them. Products that don't contain latex are not as likely to cause skin irritation or allergies, but the SDS will describe any risks from prolonged and/or repeated skin contact. Working safely requires careful application to avoid skin contact. Skin guard-type products can help avoid over exposure, but so will working carefully and controlling the application process. These skin guards can be solutions to working sloppily, but that's not the best solution. Product application should be carefully controlled to prevent skin contact with any type of curable nail coating. Also, pay close attention the SDS for your products and make sure you know which ones can cause skin damage or sensitivity and work safety to avoid skin contact with these products.

70:1 Are there differences between professional and retail nail products?

Yes, there are differences and usually those who say they are the same, are selling the retail versions and pretending they're the same. It is an effective marketing tool to suggest that a retail product is just like what you get from a salon. Often this is an exaggeration, but not always, depending on the type of product. For instance, many professional products require additional information and skills that the ordinary person does not possess, usually due to the ingredients they contain or perhaps special application skills are required. Therefore, professional training and understanding are important for many types of nail coating products. Professional products will sometimes use higher quality and/or more effective ingredients which retail products would not use, simply due to cost. Also, professional products may contain customized ingredients made especially for exclusive use in specific professional products. Such specialty ingredients are usually too expensive to use in retail products.

Even if a retail product uses the exact same ingredients, they can be very different from professional products. How? The concentrations of key ingredients in the retail product may be too low for effectiveness. Using low concentrations of specialty ingredients will save costs while giving the impression of being the same as the professional products using higher concentrations. This is why professional products often sell at a higher price point.

Of course, professional nail polish can behave and wear about the same as some retail brands. However, certain products are more concentrated and require advanced knowledge for safe use. Professional callus softeners are an example of products that are more effective, but can chemically burn the skin if used unsafely. Professional training helps to ensure proper and safe use of such products. Other professional products, such as artificial nail coating products, are an extremely sophisticated formulation and require advanced knowledge and skill. Putting such products into the retail market without adequate and proper education, warnings and other information would be irresponsible, in my view. Professional products require professional training to safely perform nail services and practice to obtain professional results.

70:2 I use a high quality certified oil, which contains bergamot. Recently a client said that bergamot could leave stains on the skin, when exposed to UV. Would you know if that is true?

I have not heard of this, but I am aware of another similar issue. The client may be confusing this with "phototoxicity". In high concentrations, some essential oils can react with UV to cause skin sensitivity and this is called phototoxicity. This is why some essential oils are recommended for use in relatively low concentrations, to avoid these types of problems. Tea Tree oil is an example. Adverse skin reactions are caused by two main factors, prolonged and repeated exposure. However, another important factor to consider is concentration or dosage. In higher concentrations, these skin damaging effects are magnified. And at low enough concentration, adverse effects can disappear completely. As the old saying goes, *"The dose makes the poison."* Everything has a safe and unsafe level of exposure. Don't exceed the safe level, and you'll be safe. One or 1000 molecules of the most dangerous poison known can't cause harm, because exposure at these very tiny concentrations is well within the safe zone even for the most dangerous substances. It is important to remember-everything has both a safe and unsafe level of exposure. Danger becomes possible only when the safe zone is exceeded and usually for a prolonged period.

According to experts, bergamot is safe when it's concentration in the product is kept at or below 0.4%. So just because the product contains bergamot doesn't mean it causes phototoxicity. Adverse skin effects are unlikely if the concentration is kept within the correct range. Many aromatherapy oils are very effective and used at low concentrations. Don't fall into the mindset that "more is better" or that "pure oil" is best, that's often NOT true with aromatherapy oils, since proper dilution can be very important for safe use. Don't be fooled by those who only have a superficial understanding of essential oils. When it comes to aromatherapy oils, many are diluted to safe concentrations and using 100% pure oil could be harmful. There is little doubt that pure, 100% bergamot is potentially harmful to skin when that skin is exposed to UV. Oddly, some have been known to apply pure bergamot to skin and intentionally expose the area to UV in a misguided attempt to treat skin psoriasis. In my view, this is a foolish home treatment and these people would be far better off seeking treatment advice from a qualified medical expert. When it comes to skin lotions, I'd expect the manufacturer of the product has likely reduced the concentration to safe levels. Cosmetic manufacturers want their products to be safe and not cause problems such as this. If in doubt, you should check the product's label and safety data sheet or SDS for information and if you still have questions, I recommend contacting the company to ask for more details.

70:3 Is it possible for air or sunlight to get into a product and reduce the effectiveness of the vitamin E by destroying it, or can it escape from the bottle? I know sunlight can affect the color, but can it break down the Vitamin E and reduce effectiveness of the product?

Vitamin E is a nutritional supplement, but when added to cosmetics, the proper name that is supposed to appear on the label is Tocopherol- since that is this ingredient's proper cosmetic name. No, tocopherol doesn't escape, yes- it is rapidly destroyed when exposed to excessive sunlight or heat. However, don't be disappointed, this is EXACTLY what tocopherol is supposed to do. Tocopherol only has one job, and it does that job very well. Heat and light exposure can cause the formation of molecular fragments called "free radicals". These are highly unpredictable, but often cause unwanted chemical and physical changes to occur, e.g. foul odors or discoloration. Tocopherol chemically reacts with and destroys these free radicals as they form in the product. However, in the process tocopherol is also slowly destroyed because it must chemically combine with the free radicals to destroy them. Free radical is a scientific name, but it exactly describes what they do. Free radicals can form freely and do some radical things to any substance they interact with. There are many types of free radicals, but they have a few things in common. They are extremely reactive and cause microscopic damage when they attack the outer surfaces of nails, hair or skin. Free radicals only exist for a short-time before they react with another molecule. Rather than react with the nails, hair or skin, tocopherol molecules react with the free radicals before they cause damage. Since tocopherol can prevent damage, it is often called an "anti-aging ingredient", but it really isn't.

This is an example of Puffery... which means to make grand statements designed to make something sound better than it really is. Puffery isn't being dishonest, but some are misled by these claims. Don't misunderstand, tocopherol does indeed help to prevent visible signs of aging, just not aging itself. It's impossible to turn back the hands of time, no matter how much the lotion or cream costs. Tocopherol should properly be called anti-damaging, but that doesn't sound nearly as good. Free radicals can develop quickly or slowly, depending on the amount of light or heat available. Increased exposure to heat or light can speed formation of free radicals. Rancidity of vegetable oils is caused by free radicals. Because tocopherol prevents rancidity, that's one of the main reasons it is added to cosmetics containing natural oils. Tocopherol will be slowly destroyed while preserving the product from rancidity, thus protecting its shelf-life. Tocopherol can help improve the shelf-life of products, but eventually, all the tocopherol will react with free radicals and will be then completely consumed. Only then will the natural oils become rancid. To preserve shelf life, you should store products in a cool location, out of direct sunlight.

71:2 I'm a nail tech and lash artist. I found a lash product that grows and darken lashes. It contains, Styrene/Acrylates/Ammonium Methacrylate copolymer. Does this means contain MMA and is this a safe product to use around the eye area?

No cosmetic product can claim to grow eyelashes. I would be very suspicious of such claims. Only medicinal drugs can make hair/lash growth claims, because this is a normal body function, which is medical. It is illegal for cosmetic products to claim they grow hair, lashes or nails, since this would be a medical claim. Only medical products can claim to prevent/treat any medical condition and cosmetics cannot. To answer your main question, this is NOT the same as MMA monomer. If it were, the label would read "methyl methacrylate" and this name would be found on the required Safety Data Sheet for the product. The term methacrylate is a category name for many different ingredients and dozens of different methacrylates are used in cosmetic products. This particular ingredient is an inert thickener that is considered safe for this application. It is a "copolymer" which means it is a polymer made by using more than one type of monomer.

Many thickeners used to control viscosity are based on methacrylate polymers, but the monomers would not be used for this purpose. There would be no reason to add any monomers to any cosmetics, other than nail coatings, so it is very unlikely you would see them. Also, monomers are very different from polymers, as I've described in other episodes. These are designer polymers formulated to have special characteristics that make them useful in cosmetics. Copolymers are NOT mixtures or blends, but instead, like the copolymer in our question, they are a new substance that has different properties and characteristics. These do not re-separate again into the original components any more than you can un-fry an egg.

71:5 What's the difference between Di-HEMA and HEMA?

These are very different, even though they have similar sounding names. Similar sounding chemical names fool a lot of people. Methanol sounds a lot like ethanol, but ingesting just a table spoon of methanol can cause blindness, while ethanol is regularly consumed by party-goers looking for fun. Both HEMA and DI-HEMA are "methacrylates" which is a sub-set of the acrylic family of chemicals. That's what the letters MA in their names stand for. However, despite some basic similarities, their chemical structures are very different, much more so than their names would suggest. HEMA is a monomer that is often used to create adhesion and is a much more likely skin allergen than Di-HEMA. Its real name is much longer, Di-HEMA Trimethylhexyl Dicarbamate. This is a very thick acrylic oligomer that often goes by another name, Urethane Dimethacrylate or UDMA. It's a commonly used acrylic ingredient in UV gels, since it cures to a hard, glass-like finish. You should be able to see that UV gels are primarily made from ingredients in the acrylic family.

72: 2 I would love to hear your opinion on nail vitamins.

If you are talking about vitamins that you ingest, I don't know of any "nail vitamin" that has convincing evidence that they work for most people. And only a very few people have ever claimed they work at all. I've talked to many who've tried them without results. It seems that much of what is claimed is based on speculation and surprisingly little scientific testing or facts. I would NOT advise anyone to purchase these types of supplements, based on this type of speculation. If these worked, we would all know it. Accolades would be piling up and dozens of other nail companies would jump on the band wagon to ride on the coat tails of the company who invented the products (at least that's what's seems happens to every other useful innovation ever introduced into the nail industry). In my view, the claims I've seen about these products are mostly clever deceptions designed to separate consumers from their money and give them little in return. However, if I ever see any convincing evidence of a nail vitamin that really works, I'll be sure to tell you about it. I probably won't have to- it will be major news when an effective nail growth formula is finally invented and proven to be effective. Don't hold your breath while you wait, as the saying goes.

72: 3 I am very allergic to Propylene Glycol as confirmed by patch testing. Am I necessarily just allergic to Glycol? A monomer liquid I am considering switching to has Glycol in the ingredients.

Propylene glycol is a pure substance, not a mixture. "Glycol" is a family name for many different and unrelated substances. The "propylene" part of this name simply indicates which individual member of the glycol family it belongs too, much like a child's name describes a specific family member. To be clear, no products contains "glycol", instead they contain alternative types of glycol, in this case propylene glycol, which is relatively safe and not likely to cause skin reactions. When people do have issues, most often it is an irritation, not an allergy. Irritations are temporary and will go away- unless the offending exposure is repeated- then symptoms temporarily return. Allergies are for life and often worsen over time, even with lower levels of exposure. Unless you were confirmed by dermatological patch testing performed by a qualified medical expert in allergy testing, you can't completely know if you are allergic. Determining allergies is difficult and there are many who provide fake allergy testing, in fact, this is a wide-spread trick used by medical quacks and con-artists.

A great website to learn more on avoiding fake allergy testing is called Quackwatch.org. This site talks about many types of phony treatments and I highly recommend you check it out. The reports I've read indicate that skin reactions due to propylene glycol are rare, even when exposed to 100% propylene glycol, it's typically used at 1-10% concentrations in cosmetics.

Again, don't be fooled by the family name "glycol". Not everybody named Schoon is related to me. So, to assume every Schoon you meet is a chemist, would likely mislead you. Nail monomers that contain the word "glycol" in their names are completely different from propylene glycol. However, this nail professional could be allergic to propylene glycol, even though most people will have no problems. This explains about how her body reacts to irritants and/or allergens. Some people's immune systems have a "hair-trigger" that can set the skin ablaze when exposed to substances that most are not sensitive toward. Anyone with these types of skin issues should proceed with caution.

73: 5 I found this on a product sold in the UK, "Warning: This product contains a chemical known to the state of California to cause cancer". Is this something to worry about?

Don't be alarmed by this label. This is a California-only warning based on a regulation called "Proposition 65". These warnings have become essentially meaningless in my view and most people just ignore these labels. You will find them EVERYWHERE in the state of California- which is home of Fear-based Activism. You'll find these signs in restaurants, hospitals, grocery stores, schools, recreational parks, literally everywhere. Here's something you should all know... aspirin, aloe vera, ethyl alcohol- the alcohol we drink- and the essentially harmless white pigment called titanium dioxide are examples of so-called dangerous substances on this overly paranoid list. This regulation started out as a "Safe Drinking Water" regulation, but it has gone widely astray. It is used by unscrupulous attorneys to sue companies over trivial issues that have nothing to do with drinking water and very little to do with safety. Remember, sunlight can cause cancer. That doesn't mean being exposed to sunlight will cause cancer. In this case, note that the label does not say the product causes cancer. I don't know what is in the product to trigger the use of this label; it could very likely be a trivial matter and no need for concern. Often these warnings are used just to keep unscrupulous attorneys from suing a company over a silly technicality. For instance, any product that contains natural, whole leaf, untreated Aloe Vera, would also be required to carry this exact same label when sold inside California, just because this substance appears on the list.

65:5 I just saw a commercial with ambulance chasers calling for a class action lawsuit over talcum powder causing ovarian cancer. Never heard that one before. Is this new?

Most of the concern is over talc that contains asbestos. Cosmetic talcs do NOT contain asbestos. A few studies have suggested possible links to ovarian cancer with any talc, but real evidence is lacking. Often these types of tests rely on a person's memory of using talc many years ago to establish the so-called "link". That's a pretty weak method which is fraught with problems and bias. It is easy to manipulate these types of test to prove just about anything. That's why this tricky method is beloved by tricksters who only pretend to do science.

Well preformed studies have NOT found an increased risk. So, I believe the known science is this: the evidence of risk to women is pretty weak. If there is an increased risk, it is pretty likely to be very small. Why worry? You can worry about anything. There is a small risk when you climb a ladder to change the light bulb. We live in a risky world, that's part of living. Talc is very widely used in many products and if this were a medium or big risk, there would be clear evidence of harm and there isn't. Even so, research continues. One thing is for sure, some will capitalize on this as a way to make money by creating unwarranted fear of talc.

67:1 I received an e-mail that said even tiny traces of the known carcinogens in cosmetics are unsafe. Is this true?

This is false and misleading information designed to frighten, not inform. Here's an example. What is acetaldehyde? It is a "naturally occurring carcinogen" found in most fruits and many vegetables. It even exists normally in "organically grown and certified" foods. So even organically grown foods also contain naturally occurring carcinogens, in fact, they contain many different carcinogens. That's because most carcinogens are naturally occurring! Even so, they don't pose a big risk because they occur in low concentrations. Concentration is the key. Don't be fooled. Many who want to frighten you, do NOT want you to realize that carcinogens don't cause cancer if the exposure is kept to low levels. This is something that many do not understand. Just because something "could" cause cancer, doesn't mean it "will" cause cancer. Exposure must be high enough and for long enough or cancer is a pretty unlikely outcome. For instance, many foods, including mushrooms and apples contain significant amounts of naturally occurring formaldehyde. And USDA Organic pineapples and tomatoes contain naturally derived 1,4-dioxane. Don't worry, these carcinogens exist in small quantities considered to be safe by scientific experts, so they are NOT expected to be harmful. So clearly, just because a substance is a "known carcinogen" or "linked to cancer", does NOT mean it will cause cancer. That's another piece of information tricksters and fear-mongers don't want you to know- they use frightening sounding claims to deceive the public. I hope we can all agree, mushrooms and pineapples don't cause cancer, even though they contain known carcinogens. Fear-based activist's groups don't want you to know these facts, so you'll never hear it from them. Why? They deceive people by telling them there is "no safe level for carcinogens." That is a load of baloney.

A substance only becomes carcinogenic when safe concentrations are exceeded and usually for prolonged periods. These substances are NOT "automatically" cancer causing and therefore dangerous, as these deceivers pretend. They demonize these substances and warn people that they must avoid them completely, but that's just their money-making-game-of-fear. Unfortunately, they've been successful fooling the public into believing the many myths they create. All people, even babies, breathe out tiny amounts of the formaldehyde. That's because our body makes and uses formaldehyde to build proteins and other beneficial substances it needs to function and live. Therefore, formaldehyde isn't automatically capable of causing cancer, and only becomes a significant cancer risk in concentrations that are many, many thousands of times higher than what was ever found in nail polish or preservatives- two sources that are often cited by these fear-mongers. Why do fear-based activists intentionally deceive people in this way? They know that by frightening people, it is easier to get money/donations. It's what floats their boat, so to speak. That's why I recommend that you never give any fear-based activist groups any money, for any reason! Because after they pay themselves, they'll use the rest to create even more deceptions and unwarranted fear. Why? That's their business model... and that's why they can't be trusted to provide the facts.

69: 1 I have two clients who've asked me if they can have artificial nails, even though they both have some form of blood cancer. Can you advise me on what products can be used on these clients as they are on medication but really want to do their nails?

In cases of serious medical health conditions like blood cancer, I would recommend they ask their treating physician for that type of advice. The physician will have their medical history and a list of all medications. They will be far better prepared to advise their patent in this regard. Some clients are forever trying to use beauty professionals as their doctor. Don't let that happen to you. Avoid giving any advice like this to any clients, unless you are a qualified medical professional. Better safe than sorry.

74:1 What is buffered acetone and is it any better or safer?

Buffered Acetone? That's just marketing puffery. So-called buffered acetone is not any safer, heathier or better. I'm glad you are asking questions and not just accepting what you are told. The obvious questions are "What does that mean" and "why is it better or good?" But many will see such claims and just assume it must be superior, because of the name. Nail techs often accept too much and don't ask these basic questions, largely because they don't know what to ask. Here are some basic questions that can be asked in order to get more details. How about, "Why is this important?" or "What do you mean by buffered?" "Why is it supposedly safer?". Don't just assume it must be good and then buy some. Ask deeper questions. And keep asking why till you get the answers you seek. Buffering means to include additives that stabilize the pH, but acetone has no pH, so the claim is total nonsense. Even if water was added and then the pH was buffered, that would be meaningless to the nail and would NOT increase safety and the added water would make the acetone less effective. Don't be fooled by buffering claims for acetone and ask questions whenever you don't understand a marketing claim.

Special topic:
Using monomer pumps vs dampen dish

Some containers are designed to pump a small amount of liquid into a reservoir in the lid. These may seem like a good idea for preventing excessive evaporation of monomer, but I don't recommend using them for this purpose. Why? The reservoir in the lid can allow excess monomer liquid to drain back into the lower container. That's not a good idea! Once the brush is used, it contains small amounts of both monomer liquid and polymer powder. The powder contains benzoyl peroxide (BPO), which is a curing agent for monomer. Containers like this can allow BPO contaminated monomer to drain from the upper reservoir back into the lower chamber of the container. As these images demonstrate, the monomer can slowly polymerize and harden over time. Even a small amount of BPO contamination can lead to pre-mature polymerization in the lower chamber and will eventually harden, as shown. Even before it completely polymerizes, the contaminated product is likely to cause an increase in service breakdown related issues for clients. That is why I recommend using a monomer dampen dish with lid. The lid has a hole in the center where a brush can be inserted to wet the brush with monomer. By placing a marble over the center hole, evaporation is easier to control. When the service is complete, any remaining monomer should be properly disposed of since it is contaminated with BPO and will polymerize.

Image 13 a & b: Monomer liquid polymerized inside dispensing pump container.
(Photo Credit: Jill Woods, Zealand Spa Salon, Commerce Twp, MI)

Image 14: Marble on top of dampen dish is a great way to control evaporation and protect the purity of liquid monomers.

Working Safely and Avoiding Skin Problems

55:4 I have a client who is allergic to water and the skin reacts even to purified water, so I can't do water manicures. Are oil manicures just as sanitary?

Oil manicures can be sanitary as well, however the client should wash their hands before the service. It is NOT possible to be allergic to water, that's not how skin allergies happen. I suspect the client may be confusing skin irritation with skin allergy, but these are very different. Skin allergies are permanent and sometimes worsen with continued exposure to whatever ingredient triggered the negative skin reaction. Continued exposure to the offending ingredient(s) can cause the skin to react and become increasingly sensitive to lower concentrations. A skin irritation will produce many of the same symptoms as allergies, so they are difficult to tell apart. Both can lead to redness, swelling, itching, throbbing, pain and other similar symptoms. Irritations go away and may never return, whereas allergies are for life. Once you become allergic to a substance, you will be allergic to that substance for the rest of your life.

56:4 How flammable is gel polish and is this why some are classified as "hazardous substance" for shipping? I must travel with my nail products and some airlines are happy for you to check these products in luggage and others aren't.

It is true that airlines are becoming increasingly concerned with what they fly as cargo. One way that nail technicians can get information about flammability and other safety information is from the products Safety Data Sheet or (SDS). The SDS will list what's called the "flashpoint" for all substances that can catch fire and burn. The flash point is exactly what it sounds like, it is the temperature at which the product will catch fire. In the US, if the flash point is below 100F (38C) the product is considered flammable. Those above this temperature are considered combustible, which means the substance is less likely to be a fire hazard. The lower the flash point is, the more flammable the substance. Acetone for instance, has a flashpoint of about 5F (-15C), which is highly flammable. This means that even if the acetone were cooled to -15C, if can still catch fire and burn. Paper is combustible, not flammable. Most don't realize that paper must be heated to more than 450F (230C) before it will burn. Combustible means the substance will burn, but it's not likely to catch fire on its own without being heated. So clearly, transporting paper is a lot safer than acetone, based on flammability. Most, but not all UV gels, have flash points that are above 100F (38C) and are therefore considered to be combustible and not flammable. Removers and solvents are usually below 100F and are therefore highly flammable. Don't be confused by the term "inflammable". Some mistakenly think this means the substance can't burn when in fact inflammable means the same thing as flammable. Anything considered inflammable also has a flash point below 100F (38C). Substances that don't burn are called "Non-flammable". Anyone who does a lot of flying may wish to consider packing the SDS sheets along with your nail products. That may help you get it checked in more easily.

56:5 In Europe, many nail technicians cut the skin with scissors or other cutting devices. What do you think about that? Some well-known nail designers are doing this.

It is true that some nail technicians do things that are incorrect and possibly unsafe. Being well known or popular is NOT the same as being well-learned or knowledgeable. There is a difference between an artist and a technician. I find that some top nail "artists" may do beautiful nails, but they are not very knowledgeable about how to be good "technicians". By that I mean that they do unsafe things in the name of "art". Cutting living skin is an example. No nail tech should be trimming or cutting living skin. That is outside the scope of a nail technician's work. As I've said before, *"just because you can bling out a nail, doesn't mean you know jack."* There are many who do beautiful nails, but do a poor job of creating nails safely. They should be admired for their artistic skills, but their poor technical skills should not be copied. It is more important to apply artificial nails properly and safely, than it is to apply them artistically. Of course, those who can do both should be admired and they are the ones who should be emulated.

56:6 What's your opinion of the UV gel polish 'steam off' machines and how safe they are in regards to the vapors being emitted and breathing them in! I like the idea of steaming off the nails, but acetone vapors sound dangerous and smelly!

I've not seen every device, but I did closely examine the original so-called "steam-off" machine. First, I will tell you that these do NOT steam-off the nails. That's just a marketing catch phrase. These devices work in a very different way- they contain a low wattage heater that warms up a small container of water. The acetone sits inside a different container and never comes in contact with the warm water or the heater, but instead floats in the warm water container. This extra warming significantly speeds removal, since warmer solvents work much faster than cooler solvents. I found the devices to be surprisingly safe and well-designed. There may be other "knockoffs" that are differently designed. What I particularly liked about the design is that the fingers are enclosed inside the container, which helps prevent solvent vapors from escaping into the air. Most of the vapors will cool, condense on the inner walls and drip back into the container. This helps to prevent inhalation of vapors and reduces the potential for accidental fires, while speeding removal. Even so, these should be used exactly as directed and should be used in a well-ventilated area to prevent the build-up of flammable vapors. Acetone should always be kept away from all sources of ignition such as cigarettes, candles, incense burners, etc.

57:4 I watched a manufacturer's instructional video which states that white spots on the nail plate are caused by dehydration. The removal technique they used was somewhat forceful. Many nail techs will use these methods with force thinking they are following the manufacturer recommendation. What do you think about this?

Manufacturers are supposed to provide fact-based information, but sometimes they provide the uninformed opinions of a few. When this does occur, if you don't complain to the manufacturer, then you become an unwitting party to the misinformation. Everyone bothered by misinformation should contact the source of the misinformation and challenge them to change it. This should happen every time, but this rarely occurs. If the incorrect information goes unchallenged, then manufacturers will assume they are correct or that no one will notice when they are sloppy with the facts. Much of the misinformation in this industry comes from private labelers placing their own label on products they don't make and often don't understand. These private labelers then feed misinformation to the nail magazines and the nail magazines reprint it as if it were facts. Where do the manufacturers get their misinformation? Often, they hire veteran nail technicians who provide them with uninformed opinions based on outdated information and myths. This may be why many nail techs have lost respect for manufacturer's directions, so they make up their own directions, which creates even more problems. If you call yourself a "manufacturer", you should be doing more research to determine the best practices and not just rely on a few favorite nail technicians' opinions.

58:5 Once I have finished applying the nail enhancement, can I ask the client to wash their hands?

Yes and this is very wise if the client is showing any sign of skin irritation, e.g. redness, swelling, itchy skin, etc. Hand washing will not cause problems as long as there is good adhesion to the nail plate. If there is a good adhesion, this will seal out the water and soap. This assumes that the nail technician does a good job cleaning and preparing the nail plate, as well as to use proper application techniques and good quality products. If this is not done, nail coating adhesion to the nail plate will be poor and washing hands can make things worse. If water and soap get underneath the nail coating, this can contribute to further lifting. But the problem isn't the hand washing, the problem is that the enhancements were poorly applied. This same will happen when the client goes home and washes their hands. Also, many nail techs are not careful to avoid skin contact with monomer liquid or UV gels, so clients' skin is often contaminated, especially on the skin surrounding the nail plate. Adverse skin reactions are caused by prolonged and/or repeated skin contact. Washing the hands after the service is a great way to ensure that prolonged skin contact doesn't occur. Then, it's up to the nail tech to ensure repeated contact doesn't occur.

58:6 Why do people put this poison on their nails? I'll never do that again after the horror I've been through. Natural is the only safe choice.

I don't know what happened to you, but generally when people have problems with nail products, it is nail technician error, not due to the products. That's why I spend so much time educating. Many nail technicians don't have the correct information and may end up damaging the client's nails. When this happens, they often will blame the products, but in my experience, it's most often caused by being under educated or just careless. Many can wear these products without damage or harm. So, natural isn't the only safe choice. I've seen damage and infections with natural nail manicures as well. Education or the lack of it is really the core of the problem. Instead, I'd modify this questioner's statement to instead say that, *"Education is the only safe choice"*.

59:4 I was told not to serve drinks in the salon to customers because "the fumes get into the drink". Is that true?

Fumes isn't the correct word, because fumes are the result of burning, e.g. candles and incenses, which release fumes into the air. Fumes are a mixture of gas vapors and soot-like particles. Nail products release vapors, not fumes. These vapors are the result of evaporation, which is where the words "vapors" is derived from. To answer your question, it would be unlikely that significant amounts of product vapors to would absorb into the drinks. I suspect that any that may be absorbed would be trivial and not an issue at all. However, there is something else to consider that is a much more likely contaminant to end up in foods and drinks. Dusts! They travel to all parts of the salon and are usually on the move with the slightest breeze or disturbance. Artificial nail dusts can settle into drinks and on foods, which is why it is generally considered to be unsanitary to serve food and drinks in places where chemical work is happening. It can be done safely, but accidents become more likely to occur. I would avoid serving food or drinks anywhere near the area where salon work is on-going. Food and nail services don't mix well.

60:3 Where can I find any information on carpal tunnel syndrome in nail technicians?

This is an important topic that is often over looked. Several years ago, as co-chair of the Nail Manufacturers Council on Safety, our group worked with an ergonomic specialist to help write an educational brochure on this subject called "Ergonomic Basics for Professionals". This information can help avoid strains and injuries caused by repetitive motion or other actions that lead to trauma or injury. This can occur when the same motions or actions are repeated over and over, since this places a strain on the body, joint, muscles, nerves, tendons, ligaments, or soft tissue. Here are some of the many tips you will find in this brochure:

1. Because problems can result from incorrect twisting, keep both the task and tools directly in front, do not favor or

lean to one side. Avoid twisting the neck, e.g. holding cell phone with the head. The head should be kept upright and shoulders relaxed.

2. Avoid reaching more than 12 inches or 30 centimeters and keep forearms parallel with the floor. This is called a "neutral position".

3. Make sure to choose a high quality, swivel chair with a seat at least one inch wider on each side than hips/thighs and adjust chair height so your thighs are parallel to the floor with your feet flat. If necessary, use a footrest to keep the feet flat and don't cross the legs or sit sideways in the chair. Make sure the chair is properly padded so that it doesn't create a pressure points on the backs of the legs or behind the knees.

4. Also, minimize other pressure points by avoiding placing arms on the sharp edge of the table. A foam tube or padding on the edge of the nail table can act as an arm or elbow rest.

5. Avoid leaning forward or backward. The head, neck, and body should face forward without twisting or hunching. Don't leaning too far forward while performing manicures or pedicures.

6. Raise and position the client's hands or legs/feet to prevent bending/stretching forward or supporting feet with your own body. A recliner chair works well and supports the client's legs in the best position for servicing.

7. Use a back rest to support the lower back.

8. Wrist and hands should be kept straight, not bent or twisted sideways.

9. When holding a client hand or finger, position it so that you can minimize the pressure from grasping. Instead, hold the finger in a relaxed manner to lessen the strain.

10. Gently move your client's hand rather than tilting your head, place a client's hand on a rest that elevates the hand to prevent forward tilt of the head and neck.

Now, there are many other useful tips in the NMC brochure, so I recommend posting it in the salon so that others may benefit from this great information. Here is a link:

http://www.schoonscientific.com/wp-content/uploads/2016/08/Ergonomic-Basics_ENG.pdf

66:2 Should I do a patch test on clients to determine if they have any sensitivities before applying nail enhancements?

I don't recommend doing a patch test for sensitivity to nail coatings and here is why. These types of products should never be intentionally applied to the skin because for some individuals, they can cause permanent and irreversible skin allergies which will likely worsen with continued exposure. Patch testing involves continuous exposure of the skin to liquid monomer of UV gels or nail polish. The problem is, doing this can be very irritating to skin and will certainly increase the chance of creating a new skin allergy that didn't previously exist. Skin allergies are caused by prolonged and/or repeated skin contact. Patch testing is considered a diagnostic medical test. It requires proper training to properly perform them but even more extensive training to properly interpret the results, e.g. to know the difference between skin irritation and allergy. There are medical allergists who are highly trained to perform these types of procedures. Therefore, if you suspect a client may be allergic, refer them to a qualified medical allergist to determine the cause of their problems. NEVER intentionally apply these types of products to skin, especially if you suspect the client could be allergic. That would be extremely irresponsible to do and could cause clients to become allergic to the products.

68: 2 When the nail plate starts to crack, is that a sign that you are allergic to the products?

No, nail plate cracking is not a symptom of allergies. Allergies are caused by the immune system, so they affect the living skin and not the existing nail plate. Typical symptoms include skin itching, redness, swelling, throbbing, warmth, tenderness or blisters. Nail plate cracking is most often due to over-filing the nail plate which leaves it thin and weaker. Also, improper removal of nail coatings can also weaken the nail plate and increase cracking. Forcible scrapping, prying and picking coatings from the nail plate can contribute to cracking. Another common reason is nail plate brittleness. Nails that are naturally brittle tend to crack more. Some make the mistake of using a nail hardener or strengthener, since these will make weak nails become stronger and more rigid. However, when applied to a brittle nail plate, these types of products can over harden to increase brittleness and cracking. The best solution for cracking nails is to reduce filing and keep the nail plate thick. I also recommend using a high quality, penetrating nail oil to increase flexibility of the plate and reduce brittleness or cracking. Regular use of these nail oils can help prevent cracking. Another solution is to keep the nail plate coated with an enhancement overlay. These can provide a lot of protection, and help keep the nail plate cracking from becoming even worse.

Contamination Control

61:2 I know you say that cutting the skin can lead to infections, but what if I only use sterilized tools? Then can I safely cut the skin around the nail?

Infections are one of the biggest problems facing the nail salon industry and they happen all the time, often because the skin barrier between us and the outside world becomes broken or compromised by injury. A skilled professional can carefully remove the visible cuticle from the nail plate, but it is risky business for anyone (other than a medical professional) to cut skin and I don't recommend this. Cutting tissue on the nail folds can lead to serious infections, even if implements are sterilized. Why? Small tears and cuts can allow infection causing bacteria to get past the skin's outer defenses to cause an infection, even many hours after leaving the salon. That is why young mothers are much more prone to infections when they cut their nail folds. Mom's get their hands in some pretty yucky, bacteria ridden things, such as diapers. This can lead to formation of large infected lumps called mucoid cysts. They can become so large they create downward pressure to distort the nail plate. Clearly, it's not a good practice to cut this skin. People have lost their fingers, hands, feet and even their life from this practice. In my view, it's is not a wise practice and should be avoided. It's not showing respect for the clients' nails or their safety.

63:3 What is the best way to choose an effective salon disinfectant? Can't I just look at a disinfectant label to see how many different kinds of microbes it can kill and chose the one that kills the most? What about bleach? Is it effective?"

Bleach is an EPA registered disinfectant. You can use concentrated, unscented bleach in salons and it is highly effective, if properly diluted (10 parts water to 1 part bleach). However, make sure it is household bleach, which contains at least 5.25% sodium hypochlorite. There is a lot of misinformation about disinfectants. Salon disinfectants tend to be very broad spectrum-which means they kill many different types of microorganisms. This question is referring to the disinfectant's label which lists the microorganisms tested to show the disinfectant's effectiveness. These are specific bacteria and fungi chosen to prove broad spectrum effectiveness. These are called "indicator organisms" and are specially chosen to be representative of many type of other organisms. Indicator organisms are usually common and some are very tough to kill, so if a disinfectant can kill them, then it has been shown to be a powerful general use disinfectant for salons. Just because a company pays more money to have more organisms tested and listed on the label, doesn't mean it must be better. That means the company paid for more testing than the EPA required. Good for them, but that doesn't mean that other disinfectants are wimpy and don't work. Quite the contrary.

If a disinfectant is EPA registered (for use as a salon disinfectant) that indicates it has passed the EPA's efficacy testing requirements, performed by an independent lab. That's not easy to accomplish. Agencies outside the US also test these disinfectants and have their own standards that must be adhered to in order to prove effectiveness. For instance, in the EU various independent laboratory tests must be passed and you will find this information on the label, e.g. Passes BS EN14476, BS EN13727, BS EN14563, BS EN13704, BS EN13624. Disinfectants can't just claim to be effective, they must prove effectiveness and meet the regulations of every country in which it is sold. For any disinfectant to be effective, it must be used exactly as directed. Salons and schools often do not properly mix, store or change disinfectant solutions and that is what can lower effectiveness. Use them correctly, and disinfectants are powerful and will help ensure the salon is a safe haven for clients. They are a necessary tool that all salons must use regularly. No salon should be without disinfectants and they should be used correctly on a regular basis.

65:2 Can ceramic bits be properly disinfected? I have heard they cannot because ceramic is a porous material, which I tend to believe. Can you confirm this for me?

The facts are, all types of e-file bits have hard surfaces and both ceramic and metal bits can be properly disinfected. Someone is "splitting hairs", when they say ceramic is more porous than metal. In terms of their ability to be disinfected, these have about the same degree of porosity. Neither is considered "porous". I would consider both to be relatively non-porous. Even so, it is important to note that neither of these are 100% non-porous. Even metals have some degree of porosity, that's how they rust. Nothing can be used that would be 100% non-porous. Marketers often misuse terms such as porous and non-porous. When you hear odd sounding claims, the two questions that ALL nail professionals should ask is, "How do you know this information is correct?" and "Where did this information come from?" The answers to these questions are usually very telling and will help you decide how believable the information is or allow you to go to the source and make sure that what you've been told is correct.

67:3 Is it possible to have a product to use in the salon to detect fungal infections? Not to diagnose them, but to make a recommendation to see their doctor, before continuing with a service.

I understand the motivation behind this question. It can be very difficult to tell a client that you suspect they have a nail infection and you can't provide a service. This idea is actually a good one and it would be great if this type of screening test existed. Unfortunately, this is not possible with today's technology. Fungal organisms are very difficult to detect and even harder to identify. Many people with fungal infections have negative tests, because even the best testing can easily over look many types of fungal organism. In the future, I expect such testing will greatly improve, but I suspect a rapid screen test for fungal organisms is still many years in the future.

68:3 A client of mine read in a beauty magazine that she should bring her own tools and autoclave them at home using her oven. Thought I'd get your scientific take on this so I could better explain this.

Beauty magazines often give foolish advice that is not based on facts, but instead based on the opinions of the author. Often the information is exaggerated. If baking in an oven would sterilize the implements, why do hospitals and dentists spend all that money on autoclaves? Autoclaves are NOT ovens. If autoclaves only got very hot like ovens, they wouldn't be very effective. Autoclaves heat up water to create high pressure steam, which is what quickly and effectively sterilizes. Heat alone cannot match the efficiency of pressurized steam. Even if the client did heat them hot enough and for long enough, which isn't likely, once the implements were removed from the oven they would no longer be sterile. Nail professionals are required to clean and disinfect or sterilize "before" an implement can be used on a client, even if that client brings their own implements. NEVER rely on the customer to do the job of ensuring the implements have been rendered safe to use. If done improperly, the client could develop an infection and the salon could be held responsible and the infection could spread to other clients. Clients may claim to have properly sterilized, but the salon has no way to verify this. Home cooking ovens should NEVER be used to sterilize salon implements. I would NOT let any client bring their own implements to the salon. It is the responsibility of all nail technicians to properly clean and disinfect or sterilize all implements before they are used on the client, and this includes those brought to the salon by clients. Most clients will not want to wait while the implements are properly disinfected or sterilized; nor will they pay for your time and effort.

So, my advice would be to refuse service to clients that insist on bringing their own tools, or implements.

71:4 I am pretty sure my UV gel manicure client has WSO. Until she sees her physician, should I remove the product? She will want them covered until then. If I continue these services or do any natural nail treatment, can this worsen the condition?

I don't think you can be sure of your medical diagnosis, because nail surface nail damage looks very much like a typical nail WSO infection, which stands for white superficial onychomychosis- a fungal infection of the nail plate. That's why nail technicians should NOT be making medical diagnosis. Regardless, this is a really tough question, here is why. I am sure you just want to help your client and that is certainly admirable. However, you should be concerned about your other clients. If you work on an infected nail, you significantly increase the risk of transmitting the infection to other clients.

I not would remove or recoat them for her. If she wants to apply nail polish herself, she can. But out of an abundance of caution, I would recommend not providing any services until the condition resolves itself and is no longer infectious, as determined by a medical professional. I know that many will ignore this advice and will remove the UV gel color before the doctor sees the nail. Those who do this should be ultra-careful to clean and disinfect everything which could become contaminated, including the table top and promptly dispose of all disposable/single-use items, by sealing them in bag that's separate from the normal trash, including nail files, cotton, table towels, etc. They'll also need to be very careful of the dust that is generated. Bacteria cannot run or jump, but they can fly by riding on a dust particle and land elsewhere to spread the infection. The last thing a salon needs is to experience a sudden outbreak of client nail infections. That's a sure way to ruin a salon's reputation with the locals, so err on the side of caution. That is my best advice.

73: 4 Can cutting the skin around the nail plate lead to infections, if the nippers or the e-file bit is sterilized and the skin is clean?

A potentially dangerous myth says that sterile implements or bits can't cause infections.

What? Of course, they can, here is how. The living skin is cut or abraded during the so-called Russian manicure or Machine manicure or E-file manicure, which is why these are considered Invasive manicures. Invasive manicures are prohibited in many places because they damage skin and make it significantly more susceptible to infections. This infection risk can last for many hours and perhaps for several days. The skin will remain susceptible to infection until the damage heals. Cutting the skin that borders the nail plate increase the clients risks of infections, even after they leave the salon. I recommend that you don't do this.

Someone told me, "Well I've never heard of that happening." Of course not! Who would openly admit they cut their clients skin around the nail plate and caused an infection? Yet I've seen this occur many times! Also, the use of an e-file to smooth, buff or abrade the skin around the nail plate is considered microdermabrasion. Many states restrict the use of the files to the nail plate only and others require special licenses, so in many places nail technicians are NOT allowed to perform these services. Check your local regulations and with your insurance company. This technique may not be covered by your insurance policy. Even calluses should NOT be completely removed from the skin due to the increased risk of infection.

The skin on the feet or palm of the hand are many times thicker than the nail folds surrounding the skin. Infections in the skin around the fingernail can quickly spread to bone to result in the amputation of a finger or hands. This is not speculation, it happens far too frequently and is a problem the nail industry must solve. Manicures should be safe and not endanger the public's health. NEVER intentionally cut or abrade the skin around the nail plate- that's trouble waiting to happen. Then it is even more foolish to place UV gel manicure products or other nail coatings directly against this damaged skin. Damaged skin is far more likely to become irritated or develop permanent allergies to nail products- More trouble waiting to happen. Don't do it! Protect the skin around your client's nails- **Don't Invade It**. Also, educate your clients about the risks of any invasive procedures and advise them against letting anyone cut or abrade this thin and sensitive tissue.

Special Topics

Episode 50 & 51: State of the Nail Industry

I'd like to overview some of the most important problems facing the nail salon industry, but I want to be clear about my intentions. My goal is NOT to criticize the nail industry, I love this industry and only want to help. Instead, I'm shining a light on the many issues that prevent the nail industry from growing and improving and offering suggestions to move the industry in the right direction. If these issues are successfully addressed, then all nail technicians, salons, clients and manufacturers will benefit. As the old saying goes, a rising tide raises all ships. These issues affect everyone, not just nail professionals. I believe the nail industry has reached an important cross-road. If the concerns raised are not effectively addressed, this could eventually result in the end of the nail industry as we know it.

I've worked as a scientist in the nail industry for more than 25 years and I know many of these are very old problems that continue to fester. However, some are new issues brought on by long-standing problems that have not been resolved and because of recent changes in the industry. Of course, the industry is growing in terms of "sales dollars", but that's not what I'm talking about. There has been an improvement in sales because the economy is finally improving, but that's to be expected and is not a sign that the nail industry in general is improving or gaining market share. In fact, I believe the opposite has occurred. In my view, the nail industry is "at best" stagnant and has been that way for several years. I know that some will disagree, but before you make up your mind, please allow me to explain why I feel this way and I'd like to offer some suggestions that I believe can help to reverse these worrisome trends.

Based on surveys I've performed over the years, I would estimate that in the US, only about 10-15% of women receive professional nail services on a regular basis. If correct, this means that 85-90% of women do not regularly go to salons for services! I suspect this is true in many other countries as well. I hear the same reasons for not going to nail salons wherever I go in the world, so these issues are universal. Of course, some women feel they can't afford salon services and would love to go if they could afford it. I'm not talking about them. My research leads me to believe that most women who could afford salon services, but don't go- is because they are afraid of nail salons. Which is really to say they are afraid of the services that nail technicians provide. Even those who receive regular nail services may trust their nail technician, but many are quick to say they are afraid of "other" salons. What is that all about? Why are so many afraid of nail salons? Some will point to unfair news or other media stories that are frightening clients.

I agree these stories frighten clients and unfairly harm salon business especially the silly and tired old stories which erroneously suggest UV nail lamps are harmful, when in fact they have been proven safe by many world leading experts who study UV effects on skin. Many of the news media stories are simply feeding the pre-existing concerns and worries which many women share about nail salons. Why is this happening? Manicures and the many innovative new nail coatings and treatments available provide wonderful services and benefits to clients. Pedicures are an awesome way to keep feet healthy and they are amazingly relaxing and enjoyable. Of course, this is ONLY true when these services are properly and safely performed by a skilled and knowledgeable nail professional. But now we come to what I believe is the root of the problem.

When a client walks in to a salon for the first time, they can NOT always expect that the services will be performed properly or safely. Why? There are many reasons, not just a few. The media has focused much of its attention on issues related to skin and nail infections, so let's start there. This issue became main stream news in 2002 when one California salon infected over 100 clients with serious leg infections that in some cases, caused permanent scarring of the legs. The infections were a clear case of negligence of the salon owner who admitted that in the two years since the salon opened, he never properly cleaned or disinfected the pedicure basins... not even once!

For more than 25 years, I have served as an expert witness in legal cases related to all types of salons- nails, hair and skin. In my experience, improper cleaning and disinfection, as well as improper sterilization are the main issue in the majority of these lawsuits. I find that a surprising number of nail technicians don't understand the difference between sanitizing and disinfection. They will often use a "sanitizer" mistakenly believing it is a "disinfectant", which can be a huge problem. Sanitizers only clean, they can NOT disinfect anything. This creates many problems. For instance, water sanitizers that are designed to be added to pedicure water while the client's feet are soaking have absolutely NO ability to disinfect the foot spa, yet some salons believe they do. I've seen this misunderstanding lead to serious client infections. Many salons don't properly use disinfectants and will make up their own cleaning and disinfection procedures, based on what's convenient for them. Some try to save expenses by short-cutting these procedures or not changing disinfectants daily or over-diluting them too much. Many times, these made-up procedures are in violation of local and federal regulations and many of them are probably not effective.

Improper cleaning and disinfection just provides a false sense of security and does not adequately protect clients. A quick spritz of alcohol on an implement or table top does nothing at all. Ineffective techniques like this only lull a nail technician into compliancy by make them feel like they are doing something to protect their clients, when they are just wasting alcohol and putting clients at unnecessary risk. If cleaning and disinfection are properly performed, these procedures are extraordinarily effective, and clients would be protected and salon-related infections would be very rare. Problems arise when these procedures are not properly performed. For instance, some salons use so-called UV sterilizers, which do not properly sterilize salon implements. In my view, these are a waste of money. These devices create a greater risk for clients. Some salons have switched to autoclaves for sterilizing small tools and implements, but most don't properly maintain these devices and never perform the monthly spore testing that is required to ensure the autoclave is working properly. Hospitals do these tests weekly, because they know autoclaves can appear to be working, but are not properly sterilizing, which is why they must be checked regularly with spore testing. Nail and skin infections occur too often as a result of salon services and this is a main reason so many doctors warn their patients to avoid going to nail salons. This information should be the first thing nail professionals are taught and it should be a main emphasis throughout their training, but too often cleaning and disinfection is not a focus.

This brings up the next issue, nail technician training. There is an interesting new problem facing the nail industry called "nail art"? What? How can nail art be a problem? Don't misunderstand, I love nail art and I'm amazed by the talent of many nail artists. So, what's the problem? Many new nail technicians are being drawn into the field because they want to artistically express themselves through nail art. Nothing wrong with that, if properly channeled. However, this trend has a dark side that must be addressed. Too many are overly focused on nail art and they lack even a basic understanding of the natural nail or how the products work or how to use them safely. Many nail instructors tell me that their students now only want to learn how to do nail art and they don't care about learning anything else. Also, many nail artists have resorted to using non-cosmetic glitters and other potentially unsafe colorants NOT designed or intended for cosmetic use. When the nails are filed, the nail technician is exposed to the dusts of these non-cosmetic colorants and this can lead to adverse skin reactions and other health risks. Nail technicians must learn much more than how to do nail art, yet many are overly focused on nail art and do not have the other skills and knowledge needed to work safely. That's why I say the nail industry has too many nail artists, not enough nail technicians. Just because you can Bling out a nail, doesn't mean you know Jack. In other words, being good at nail art doesn't mean you're a good nail technician, just like being good at decorating a cake doesn't make you a good baker. Don't fool yourself, nail art is just the icing on the cake, it's not the cake.

It is my belief, the biggest issue in the nail industry is that very large numbers of nail technicians do not have enough information or knowledge to safely or correctly perform their services. Yes, there are a great many highly knowledgably nail professionals in every country and I applaud them. Even so, on a world-wide basis, nail education is greatly lacking, and the level of knowledge is below what it should be to safely and correctly use nail products. No country is exempt from this issue, and that includes the US and other countries that require nail professionals to be trained and licensed. Much of what some consider to be "nail education" is better described as a confusing collection of misinformation, myths, truths, half-truths and deceptions. Most nail techs know very little about the natural nail, which alone should be a concern. Many confuse the nail plate with the nail bed and in my experience most nail technicians can't name the parts of the nail nor do they understand the basic structure or function of each part of the nail nor do they understand how these parts work together to create a healthy nail.

Would you take your car to a mechanic who didn't know the difference between a transmission and the engine? Probably not! To me, having a deep understanding of the nail is basic and fundamental to being a nail professional. Yet many focus only on application skills, but use improper removal or other techniques that overly damage the nail not even realizing how to prevent nail damage. Instead, too many in the industry just accept nail damage as something normal and unavoidable. When the surface damage becomes too noticeable to ignore, they'll tell their client they have "dry nails" and need to use nail oil when the facts are- nail oils only lessen the appearance of many types of nail surface damage, not repair the damage. Other than using scraping or other forceful removal methods, one of the most damaging techniques that nail professionals use is to over-file the natural nail. Many don't understand the consequences related to over-filing. Nail technicians that don't understand these basic points are far more likely to damage the nail than they are to protect and nurture. Each time a client's nails are needlessly damaged or made overly thin and weak, that's another black eye for the entire nail industry.

Many nail techs confuse "education" to mean application skills. Application skills are only a small part of what nail technicians need to know. Most nail technicians focus their time developing application skills, so they become quite adept at using the products. Then many make a critical error, they begin to think they know enough. Many will intentionally shun any opportunity for advancing their education. They don't take classes, don't read trade magazines (maybe look at the pictures) they know enough already- or so they think! I have the honor of knowing many of the finest nail technicians in the world, and I would say there are very few who can truthfully claim they know "enough" about the nail and nail products. They would never claim to know it all, because they have a thirst for more knowledge. Do you think that's how they got to where they are? Or was it by realizing they don't know it all and likely never will and not letting that stop them from trying to learn more?

Many nail technicians are uncomfortable with learning anything that might challenge what they were taught by their mentor and instead would rather ignore the facts. Some rely on the Internet for information, but this has actually become a convenient way to share misinformation on a grand scale. Now misinformation flows like water, it's everywhere making it even harder to distinguish between facts and myths. Some veteran nail technicians use the Internet to teach their bad habits to new nail technicians who don't know any better. Some teach and encourage nail technicians to work outside the scope of their jobs.

What do I mean by that? Nail professionals should not diagnose medical conditions, yet some incorrectly teach others to diagnose and identify an infection by the nail plate's color or appearance. This is extremely difficult to do, even for trained medical professionals, which is why a wise doctor will send a sample of the infection to a laboratory for identification. Many nail professionals don't stop there, they attempt to tell the client how to get rid of the infection, without realizing they are actually prescribing treatment for a medical condition.

That's clearly outside the scope of a nail technician's work and against the federal laws of most countries who require a medical license to treat medical conditions, like nail infections. Only qualified medical doctors should be diagnosing and identifying treatment for infections or any other medical condition of the nails. Telling clients they have an infection and then directing them to soak in vinegar is an example of improper diagnosis and treatment of a medical condition, the same is true when nail technicians give clients an antibacterial nail oil to treat the infection. All of this is outside the scope of a nail professional's work and could delay a client from seeking a correct diagnosis and proper treatment by a medical professional. Nail professional should NEVER diagnose, treat or prescribe a treatment for any medical condition. Nor should they intentionally perform any potentially unsafe medical procedures.

I'll give you an example of an unsafe medical procedure. Some nail professionals attempt to apply artificial nail coatings to finger or toe nails that are missing all or part of the nail plate. All artificial nail coatings should be kept off the skin and NONE are safe for nail technicians to use to replace missing nail plates. To do so requires the nail professional to intentionally expose the sensitive living tissue of the nail bed to artificial nail coatings that are only safe for application to the nail plate. Creating a prosthetic nail is a medical procedure that requires highly specialized training and caution, yet some are teaching and practicing these improper procedures. All nail manufacturers should be instructing nail professionals to always avoid contact with the skin, due to the potential for adverse skin reactions. I know of no exceptions to the rule, even though some companies are improperly and incorrectly claiming their products can be used for these applications by nail technicians. As I've talked about before, one important way to avoid developing a permanent allergy to the nail coating products is to never touch them to the skin, until they are properly cured. Applying artificial nail products designed for the nail plate to the living skin of the nail bed is a form of product misuse that can lead to permanent allergic skin reaction which worsens with each additional exposure.

Here's a very real problem that concerns me greatly even when this type of correct information is discussed on the Internet, it's been my observation that nail technicians with the correct information are often shouted down by those who are incorrect and refuse to let go of their misguided beliefs.

The Internet has become the place for some misinformed nail technicians to argue over who's right, when both sides are often wrong. This is probably why so many are so easily confused or fooled by deceptive marketing. However, a nail professional with fact-based knowledge is more difficult to deceive or mislead. That's why it's said that information is power. The information in this book and my Internet video series give nail professionals power over the nail, power over their products and mostly power over health and safety for themselves and their clients. There are some who don't want nail technicians to become too well informed. Why? An educated nail professional is harder to fool. Of course, most manufacturers are honest and try to do a good job... I'm not talking about them. I give those companies special kudos if they are making strong efforts to be responsible and fair marketers and to educate the proper use of their products. But there will be those who don't play by the rules and their business model seems to be aimed at taking advantage of others who don't understand the facts. For instance, some UV lamp manufacturers and distributors would find it MUCH more difficult to fool nail technicians into buying the incorrect UV nail lamps "IF" most nail professionals understood that using an incorrect lamp not only increases the chance of service breakdown, this can cause nail technicians to develop a permanent allergy to their favorite nail coating products. Using an incorrect UV nail lamp is a leading cause of adverse skin reactions for nail technicians who use UV gels and this can lead to damage of client's nails, as I've discussed many times in this book series.

No manufacturer of any generic type of nail lamp can accurately claim their nail lamp properly cures any UV gel. They can't know, since they can't have tested all UV gels. Besides, it is both chemically and physically impossible for a one nail lamp to cure all UV gels for the reasons I've explained. This is called Cherry Picking! If you've not heard the term, some think of it as picking out their favorite products and using them in combination with other favorite brands. Essentially, cherry picking is using together different products that were made by different manufacturers and NOT designed or intended to be used together. This is a form of product misuse that can lead to service breakdown, adverse skin reactions, as well as, nail damage, even nail infections. It is also a problem when a manufacturer develops a product that requires nail technicians to misuse other brands of products. For example, if a company developed a nail polish and instructed users to mix it with any brand of UV gel, I don't think that is right nor is this safe to do! This leads to incorrect curing and increases the risks of adverse skin reactions and by doing so, this encourages the misuse of other company's brands of UV gel. Cherry Picking just encourages nail professionals to take it upon themselves to change the manufacturer's directions and use products "their way" which many times leads to problems in the salon.

Some sell polymer powder that's to be used with any other manufacturer's monomer liquid? What? That's risky business in my view, because this can be unsafe for those who must use these products on a regular basis. Most powders are a blend of methyl and ethyl methacrylate, but that does NOT mean they are the same, there are many important differences between various brands of powder even though some try to mislead nail technicians into believing "a nail powder is a nail powder and they are all the same", which is incorrect. I could make one-hundred different powders just by blending the ingredients in different ways and at different ratios each blend would cure differently and produce nails of very different quality, strength, and durability. Improperly cured monomer liquids are more prone to cause allergic reactions. That's why it is a mistake to use the incorrect powder with a monomer liquid. The liquid and powder are a matched set and should be used as such to ensure a proper cure and they MUST be used at the correct ratio of liquid to powder to create a medium dry bead.

The image below provides visual guidance that can help nail technician know if their mix ratio of monomer to polymer is correct. If too much polymer powder is used, the bead will be too dry and it will retain its shape without flowing, as shown. If the bead contains too much monomer liquid, the bead will lose its shape and drop more than 50% of its original height and will often form a ring of monomer liquid around the base of the bead. A medium dry bead is best, as shown in the middle. This bead retained its shape, didn't settle more than 25% of its original height, and no ring of monomer liquid formed around the base. This bead has just the right amount of monomer to polymer.

Image 15: Beads of acrylic monomer liquid and powder at different mix ratios.

The vast majority of the problems in the nail industry occur because so many nail technicians are not taught how to properly use nail products and instead are taught incorrect and damaging techniques, designed to save time not to better serve the client or to protect the health of their nails. Some of these damaging, but time saving techniques, come from nail veterans who think they've come up with a "better way" to use the products and they encourage others to misuse the products as well, often without understanding the problems this can cause.

Unless a person has a deep understanding about the chemistry of nail products and the nail, it is risky to disregard manufacturer's directions and warnings. Making up your own blends and instructions can significantly increase the risks for skin allergy, infections, or nail damage. I think it would be helpful if all nail manufacturers played a greater role teaching safe use of their brands and product lines, as well as to promote a greater awareness and understanding about safe use and practices related to the use of nail products, in general. Ensuring the safety of the client and the nail technician through education is important to the future of the nail industry. Some manufacturers do a great job, while other do not and continue to propagate misinformation from the past. Many are only private labelers who have little technical understanding of the products they sell. For example, some promote over filing of the nail plate or improper removal techniques that damage the surface of the nail plate.

All of this information should be taught in nail courses, but many instructors are not keeping up with new information and are not focused on safe and correct use. Instead, too many are teaching their misinformed opinions and their own bad habits to students. For instance, a common problem related to over filing the natural nail is too use too much downward pressure or too low of a grit nail file. A 180-grit file can very quickly do a lot of damage to the natural nail when used too aggressively or held at an incorrect angle. An electric file can do even more damage in the hands of someone without proper training. Over filing just thins and weakens the nail plate and can make it more difficult to remove the nail coating without damaging the natural nail.

Also, some nail veterans teach classes that fool people into thinking they have enough skill and information to do nails after only a few days of training. Excuse me! But there is NO WAY anyone can learn enough to safely and properly perform these services with just a few days or a week of training. Those irresponsible trainers are creating a huge problem for the nail industry, since many of these students go on to damage their client's nails or cause allergies because they lack proper training and information.

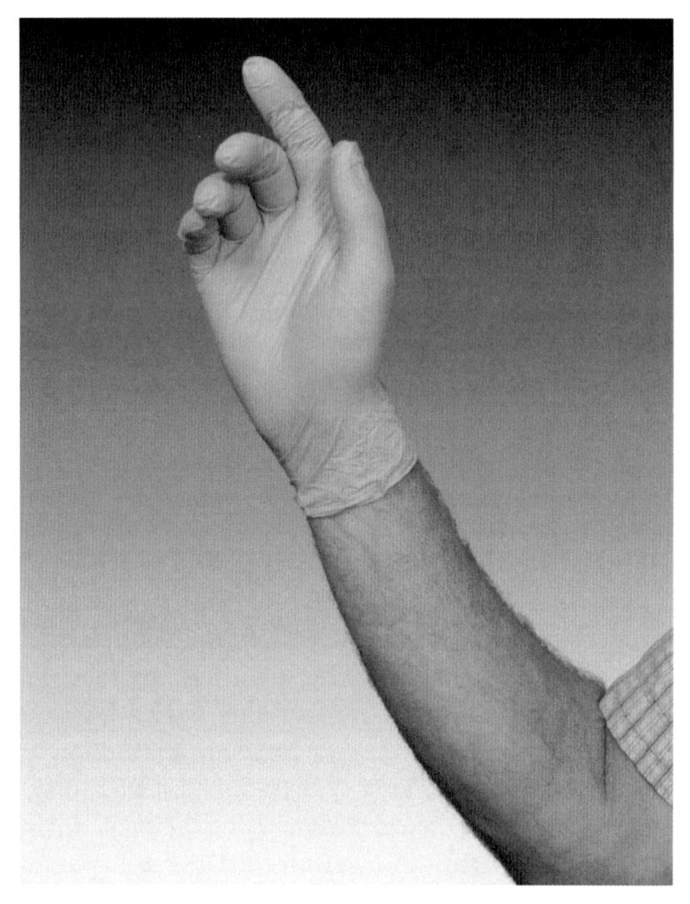

Image 16: Wearing 8 mil, disposable nitrile gloves offers the best protection for nail technicians.

The final issue I'd like to discuss is this, too many salons are not properly ventilated and there is too little focus on safety. The nail industry in general is in denial about the use of ventilation and gloves. These are important safety tools that every salon needs. Salons and even mobile technicians need to invest in proper ventilation that is appropriate for the work being performed. Also, nail technicians should wear disposable nitrile gloves to help prevent skin contact. These issues are very real problems that must be solved for the nail industry to grow and prosper. Everyone must play a part, not just nail technicians. Those who call themselves "manufacturers" need to become more informed and aware of these many issues and need to seek solutions, not compound the problems, rather than turn a blind eye toward these problems and accept them. One small manufacturer said to me recently, "Hey, it is what it is, just accept it". No, we should not accept this as "just the way it is". We should collectively work together to change the industry for the better. Please, if you care about the future of the nail industry, get involved and do your part to correct these many issues, you'll be helping the entire nail industry if you do. Remember, a rising tide raises all ships.

Understanding Salon Activist Groups

53:1 Why is it that every time you talk about activist groups for cosmetics, you call them "fear-based" and disagree with just about everything they say. Somebody from one of these groups gave a talk at my school and she said things that make a lot of sense to me. I don't know why you don't work together. Have you ever tried to talk to them?

The person asking this question will be surprised to learn that I started working with cosmetic activist groups about 15 years ago. During that time, I've learned a lot about how they operate, as well as their true motives and ultimate goals. Actually, I should say I've been trying to work with these groups, but I've not been very successful. I reached out to several of these groups when they first formed and offered to work directly with them. I've been to several meetings which they graciously invited me to attend and were very nice, but they made it clear they had very little interest in working with me, because I don't agree with their tactics, methods, or information. I believe I could help them to see a better way to improve safety in the salon industry. They don't need to attack with misinformation and cleverly worded scare tactics based on distortions of the facts. These have become their weapon of choice, and they don't seem to want to abandon these destructive methods. They don't have a proper understanding of the facts and are closed minded. They ignore any information they don't like, even when confronted with facts. I know, I've tried and not gotten very far by explaining facts they aren't interested in hearing.

I disagree with their use of irrational fear and deceptive tactics to force a change in thinking or to change rules and regulations. The ends don't justify the means, and these misguided tactics needlessly frighten people and create unwarranted alarm. Even so, I do agree with these groups on many important points. By working together, we could achieve the goal of making salons safer and ensure safe working conditions for all salon workers. If that were their real goal. Here is the difference in our approaches to make salons safer. Fear- based activist groups want the power to demonize and ban products, services and ingredients so they can eventually eliminate certain types of services, such as liquid and powder and UV curing artificial nails. I don't agree with this approach. Instead, I believe the evidence shows these ingredients, products and services are safe when used properly, so the goal should be to teach professionals how to work properly and safely, not to take these things away. We should do a better job teaching nail technicians how to work safely. That's what's missing. We all know that automobiles can be dangerous, but we don't ban them. Instead, we teach people how to drive more safely and to look both ways before crossing the street. It seems to me that these groups have already made up their minds and don't want to listen to the facts. They are used to doing all the talking and don't want to listen. Why? Because they fear the facts. That's why they refuse to listen to the knowledgeable scientists in the nail industry and try to keep us out of every discussion where salon safety is involved.

Once, I asked one such a group why they keep refusing to work with me or any other scientists in the industry, after all, we want the same thing... to make salons safer. Their honest reply caught me off guard, and at first I didn't know what to say... and I'm rarely speechless. I was told this, *"No, we don't want the same thing. You want to teach nail technicians how to work safely with toxins. We don't want to teach them that, we want to eliminate all of the toxins."* What could I say to that? Don't they believe in safety training or use of ventilation or other safety equipment? They see these as an excuse to allow nail technicians to be exposed to so-called toxins- which is pretty silly if you think about it. These fear-based activist groups paint everything in black and white. They label chemicals as "Good or bad" and bad means zero tolerance. As I was once told there is no safe level for any toxic substance, which is a ludicrous statement based on a lack of knowledge. Ingesting excessive amounts of salt over a short period would be very harmful, but swimming in salt water is not harmful and swallowing a little from time to time is perfectly safe. I know, I surf and have swallowed lots of salt water and I'm fine. Many things that are toxic in high concentration are safe at lower concentrations. But, the goal of these groups is the demonize substances and force their elimination.

My approach is different, I'll teach you how to work safely and give you the facts and knowledge so you can avoid the risks. Even pure water can be risky if used incorrectly. Stick your head in a bucket of pure mountain spring water and you'll find this deadly chemical kills in just a few minutes! Is water scary stuff that should be banned? Just because water is linked to the death of millions and we know it can kill, doesn't mean it that it will. What "can" happen is very different from what "will" happen. It's all about how water is used, right? Some aromatherapy oils contain traces of naturally occurring carcinogens, as do many natural substances apples, tomatoes, lobster, and dozens of other foods. Does that mean these aromatherapy oils and apples cause cancer? Of course not, that would be extremely unlikely, but not "impossible". That's the game these groups play, they live in a world where they fear everything that "could happen". Scientifically speaking, I can't say there is "zero risk" from using aromatherapy oils, because of the tiny traces of a carcinogen, even though cancer is extremely unlikely at such small concentrations. The risk may be trivial and not important to consider, but no scientist will say the risk is zero! That's a favorite trick often used by fear-based activists, they often fool the public by getting experts to admit there is a "risk". Knowing the risks can never be zero. An asteroid could hit our planet while you read this, but it's not very likely at all and I suspect you are not concerned with the minuscule possibly that it "could happen"! The risks are NOT zero, but pretty close to zero. We live in a risky world and we shouldn't waste time worrying about unlikely risks. It's obvious that using aromatherapy oil is probably a lot safer than riding a bicycle and we give bicycles to children. Bicycles are linked to death, brain damage and paralysis! Instead of forbidding bicycles, we teach children how to safely ride their bike, make them wear their helmets and obey some basic rules of bicycle riding. Isn't the issue more about HOW we work with products? In most cases, that's more important than which products we use? We must stop this fear-based silliness or before we know it 10-free nail polish won't be safe enough.

These groups have created a tangled web of distortions and misleading reports that can't be substantiated with fact-based information. So, when faced with the facts, they just refuse to listen. They would rather cherry pick one or two poorly performed studies based on junk science contrary to the majority of existing scientific evidence. They use these ridiculous studies to dupe the media and well-meaning politicians, both are favorite targets. Politicians and the media have the power to change and/or control the cosmetic industry and individual businesses, including salons. And that's the real goal of these groups, they want to tell manufacturers what they can and can't make. They want to control what distributors sell and most importantly, they want to control what products you can or can't use in the salon. They think they know best, but how could they? They refuse to work with the most knowledgeable scientific experts in the industry? Instead, they handpick their own experts who agree with them and then they provide money to fund their research. If their research turns out to not agree with what they are saying, these groups just bury the information and never let it be published.

Another favorite tactic is to exaggerate extremely small risks and blow them hugely out of proportion. In this way, they can fool people into fearing something that is extremely unlikely to occur. It is a fact that golfers are sometimes killed by lighting while they are golfing and this continues to happen. What does this information mean? People shouldn't golf because they could be killed by lightening? Or to be on the safe side, golfers should avoid golfing if there is any chance of lighting, otherwise, golf away! What's the difference between these two? These different approaches create very different perceptions about golfing. The first one says that golfing is to far too dangerous because there have been deaths by electrocution. That doesn't make any sense if you think about, which is why these groups don't want you to think. They don't want you to know too much... you might start thinking logically. To continue the analogy, a fear-based group would say, *"We can't wait for more studies; golf should be banned for public safety reasons. Not another innocent victim should die while practicing the clearly dangerous sport."* Or perhaps we should take a more reasonable approach, such as to say, *"Golfing is safe, but you need to follow some basics rules of safety, e.g. discontinue golfing if there is any chance of lightening and check your local weather before golfing. Otherwise, have fun and please golf safely."*

This example illustrates how easy it is to twist the facts to create irrational fear and use that fear to motivate others into doing what you want them to do. So, the next time you get an e-mail about the latest toxic scare and a request to "donate money", don't do it! They will just use the money to needlessly frighten your friends and customers. And if you read or hear any of this fear based nonsense about nails or nail products please ask them. Are you working with the leading scientists in the nail industry, if not then why?

Spotting Myths/Misinformation

54:4 I do my research, but there is so much incorrect information coming from other nail technicians and even from manufacturers, how am I supposed to separate the facts from the myths?

Determining fact from fiction is an age-old problem that has existed since the beginning of civilization. In other words, misinformation has always been a problem. The best solution to this dilemma was discovered about 2500 years ago by a Greek philosopher named Socrates. Socrates and the people living in his time didn't have to deal with the all the baloney circulated on the Internet or by the news media, but they sure did have their own fair share of myths and misinformation to deal with. So, Socrates developed a method to help him get the facts and he taught this method to his followers. His method was so extremely successful that people regularly use it to this day! This method is used by judges, lawyers, reporters, detectives and scientists looking for facts and clues about the facts, but anyone can use this method because it is very simple, yet highly effective. Socrates's method can be boiled down to just these two words, "Question Answers". Yes, it is that simple, yet it is powerful! Don't just accept what you are told. That is how nail technicians are often fooled. Instead, ask lots of questions, and keep asking until you are satisfied you fully understand. It's all about the questions you ask. Here are some examples for you to try.

The best way to get answers is to ask open ended questions. Start with the simplest question of all... Ask "Why"? When someone tells you something you suspect is incorrect, ask them this question, *"Why do you believe this is true?"* And listen carefully to their answer for clues about what they are really saying. Or ask

"How do you know this?" Another interesting question is, *"Where does your information come from?"* and *"What does this mean?"* Or my personal favorite, *"Please help me understand why I should believe this?"* Or, *"Why is this information important?"* If you don't understand their explanation say, *"Explain that to me in a different way so that I can understand."* Or *"What does that mean?"* These are just a few examples of the type of open-ended questions that can help you to do better research and get a deeper understanding about any issue.

Those who don't know what they are talking about will have a hard time answering such questions. Those who want to deceive you, will not like being questioned. In either case, if someone doesn't know the facts, it will be obvious when these types of questions are asked. And those who don't want you to know the facts, will become uncomfortable. When people ask me questions like this, I love it! This provides an opportunity to explain in greater detail and lets me know the questioner is interested. It's an opportunity to provide more detailed information, which I always appreciate. Don't stop asking "why" and "how" until you get to the root of the issue and feel you understand the facts. You'll learn that some don't know why they believe, they just do and will have no clear answer to your questions.

Finally, don't just accept what you hear or read and don't just look at the headlines and expect you know the facts. Headlines are often intended to fool people, since those writing the headlines know most won't take the time to read the article and won't ask questions. Many are afraid to ask questions, but now you understand the tricks used by professionals who ask questions for a living. With practice, you'll find it's easy. Take the time to Question Answers! And you'll be the wiser for it. Like anything, it takes practice- so dig in and don't give up until you get the answers to your questions.

Secrets to Innovation

59: 1 What is the most important thing for someone to consider when developing a new innovation?

Great question with a simple answer, *"Never fall too in love with your own ideas"*. Often people will get a great idea and love it so much, they won't listen to input from others. They think their idea is so good, it doesn't need to be changed and couldn't possibly be better. That's unlikely. In my 25+ years in the industry, I've never seen an innovative idea that couldn't be improved upon. Others will view your idea in a different light and may see unexpected opportunities or roadblocks that could easily be solved to improve the idea. Sometimes, these unresolved issues create opportunity for competitors to create a superior product or solution. That's one reason why problems/issues should never be ignored so they become roadblocks. I always ask those testing my new ideas to tell me what they "don't like", rather than what they like. What they don't like is called "opportunity" to improve.

So, to all of you innovators out there, don't be so afraid someone will steal your ideas that you won't share them and instead keep them hidden. This can be paralyzing! Instead, seek out the input of those you trust and listen carefully to what they say. Ask what they don't like, what they would change or like to see done differently... and ask why. Listen carefully, be respectful of the input and then take suggestions, concerns, and complaints very seriously and make any necessary changes, that is, if you hope to make your idea the best it can be. Good luck and Happy Innovating!

Organic Nail Enhancement Claims

60:2 How can I tell if a nail enhancement is organic?

This is easy to determine, yet many are very confused by this issue. Largely because many marketers use this word incorrectly. The word organic is used to describe so many different things, that it has now lost its meaning. It certainly does NOT mean what most assume it means. Here are the facts: All nail enhancement coatings are based on organic ingredients and there are no exceptions. What do I mean by that? Organic substances are made using the element carbon. Since living things are all based on carbon, all living things are organic and so is their food. Even bacteria, fungus and viruses are organic. So just about everything on the planet is organic; except for air, metals, minerals, and water. Even petroleum is based on carbon, so road tar and gasoline are both "organic". Being organic is no claim to fame. Many confuse the term "organic" with other established terms such as "organically grown" or with "certified organic". These are VERY different, and it is a BIG mistake to confuse them, yet the public doesn't understand the difference. The public incorrectly believes that "organic" means "safer", when it does not. Organically grown peanuts could kill a person with a peanut allergy. Organic standards specify how foods are grown and handled, NOT how safe a product is. That is a big deception. These standards do NOT ensure safety, despite the many over bloated promises and false claims that some marketers use to fool the public. It's a fact that even organically grown foods can also be unsafe and this has occurred in the past.

To date, no artificial nail coating is based on organically grown ingredients and none are certified organic by any recognized agency such as the USDA or Ecocert. This demonstrates that no type of artificial nails are any more organic than another. This is a meaningless claim often used by tricksters attempting to fool nail professionals into using their products. My advice, don't purchase any artificial nail coating because you believe it is certified organic and I recommend that you be automatically suspicious when and if you hear such claims. Avoid companies that use such deceptions. Otherwise, you are rewarding them for misleading you and they'll be back for more of your money.

Pregnancy in the Salon

60: 6 I'm a nail technician and now I'm pregnant. In Greece, some doctors believe that the smell of nail products is dangerous for a fetus and even pregnant clients shouldn't have their nails done. Is this correct?

Doctors should have evidence for what they believe. So, what is the evidence that using these products harms unborn children? I don't know of any evidence, in fact the evidence demonstrates this is not correct and these are indeed safe. Which evidence? Well, just for starters, many millions of people use these products every year and pregnancy related problems have not been shown to be a problem over the last 30+ years. So why would these doctors believe that this is an issue? What is the basis for their concern? Is it just the funny smell that seems strange to them? My guess would be that these doctors know little to nothing about the chemistry or safety of nail coating products, yet that doesn't stop them from pretending they do. This is fairly common. Instead they are probably just repeating what they've heard others say, or they are forming their opinions based on the odor of the products and not the facts. Many doctors express their personal opinions and pretend that they are backed by facts.

Based on all the information I have read on this issue and my understanding of the chemicals used and the way in which they are handled, I do not believe there is evidence to demonstrate nail technicians are at greater risk for pregnancy issues/complications related to their work. I believe that if a nail professional works in a safe fashion, using their products safely and correctly, to include working in a well-ventilated salon, there is little reason to be concerned that their work will harm their unborn children. There is an old saying in Holland, "Fear gives you the worst advice". That's why it is best to base your decisions on the facts- not fear! Note: For more information on this subject see Volume I, Special Topic 4 page 133-135.

Understanding Natural Oils

62:1 What is so different about argan oil that its called a "miracle oil", is this just hype or is this oil really different from other natural oils.

I want to provide a deeper understanding of vegetable oils used in cosmetics, including argan oil. Botanists are plant scientists and they define fruits and vegetables like this; fruits are the seed-bearing part of a plant and vegetables are all the other parts such as the seeds, roots, leaves, bark, stems, etc. Cosmetic scientists used a broader definition. Cosmetic vegetable oils can be extracted from any part of the plant and are considered vegetable oils, regardless of the plant part used. Examples of vegetable oils extracted from fruits are olive oils, avocado oil and palm oil. Examples of oils extracted from nuts are almond oil, macadamia nut oil and coconut oil. Many oils are extracts of seeds, such as canola oil, grapeseed oil, palm kernel oil and argan oil.

These oils have important physical properties that determine their usefulness in cosmetics. For example, the melting/solidification temperature, evaporation rate, color, odor, viscosity (thickness), shelf-life stability, solubility and the ability to penetrate into nails, hair or skin. These properties are all determined by the main components of the oil. Most vegetable oils are practically non-evaporating liquids, while some are semi-solids. Vegetable oils are usually in a liquid state at room temperature, but may begin to solidify when cooled below room temperature. Those that remain solid at room temperature are referred to as "vegetable fats" or vegetable waxes and these are not the subject of this discussion, I'm only discussing the vegetable oils. Some vegetable oils are said to be volatile, which just means they evaporate very quickly. These are often called essential oils, because they contain the "essence" of fragrance of the plant. These are removed from vegetable oils and sold separately. Essential oils are not the subject of this discussion.

There is a lot that can be said, but so that you may better understand cosmetic vegetable oils, I'll share some general rules-of-thumb that can guide your understanding and future investigations. For instance, it is important to understand that when our noses smell essential oil vapors, we are inhaling a small sampling of the millions of essential oil molecules that are floating in the air. A good rule to remember is; in general, the smaller a molecule, the more easily it can evaporate. How do scientists know the size of a molecule? We look at its molecular weight. When it comes to many things, including vegetable oils, the lower the molecular weight, the smaller the molecule and the more easily it can evaporate. Therefore, the molecular weight is a good indication of the evaporation rate. Also, the smaller the molecule, the more easily it could penetrate the nail plate, skin or hair. But it would be overly simplistic to just think about the size and weight of the molecule. The shape can have an even greater influence than size. Molecules with a bulky shape are often blocked from penetrating the surface of nails, skin or hair. However, molecules with a similar composition, but a sleeker shape can pass by the surface and enter. As I mentioned before, essential oils are removed from vegetable oils, which helps explain why cosmetic vegetable oils have very little odor, at all. Some describe them as slightly "lard-like". Olive or safflower oil are good examples of the typical odor of vegetable oils. A lot of people tell me they love the smell of jojoba or argan oil or sweet almond. No, they don't, when a vegetable oil smells great, it is because essential oils and/or synthetic fragrances have been added back into the oil. Nothing wrong with that. There are many awesome products that do this. But, the users actually like the product's overall formulation, including the added fragrance and not the specific oil which is probably pretty boring.

So why are these vegetable oils useful in cosmetics? I've asked over a hundred people this same question and their answers are summed up as, *"oils soften, condition, moisturize, hydrate, repair and heal"*. Soften and condition are certainly correct- and these are a major benefit. Moisturize and hydrate are misunderstood, but is sort of correct. Vegetable oils only slow down the rate of water evaporation from the skin or nails. These oils contain NO moisture, so they can't hydrate, but instead they slow evaporation and increase the number of water molecules remaining in the skin. Repairs? That's pretty unlikely. I don't see how any vegetable oil can repair damaged hair, nails or skin. So, I would call this a false claim. These oils can hide surface damage, but they can NOT repair the damage. Nor can they repair split ends. When the oil wears or is washed off, the damage remains. However, these oils can help prevent existing damage from worsening and can increase resistance to future damage- so that's a great benefit. Can they "heal"? No way! That's not a proper cosmetic claim and no one should be making such claims about any cosmetic. No cosmetic can claim to prevent or heal any disease, and this includes infections. Cosmetics don't prevent or treat infections and can't make such claims. Not only are such claims false, they are illegal to make. Why are these plant-derived vegetable oils so misunderstood?

This is just my view, but I think there is too much marketing hype around the "name" of the oil. Little is said about the composition of the oil, and instead the focus is on the oil's name. Think about it. Why is the "name" of the vegetable oil important, except for marketing purposes? The name is used as a marketing hook, to get consumers interested, so I get it. Many marketers tend to focus on oils with cool names from exotic locations. That may be a lot of fun, however it's not important to a product's function. Many over-promise what the oil can really do, some act as if they are almost magical! Marketing aside, why are these vegetable oils used at all? After the essential oils and vegetable fats and waxes are removed, what's left are some extremely useful substances called, "fatty acids". Fatty acids is the chemical family these substances belong too. Other useful substances are found at much lower concentrations, but fatty acids are the majority of the vegetable oils. Fatty acids is not a nice sounding name, so it has no marketing value. You'll never hear about them in an advertisement. They are NOT the same as fat molecules and they won't burn the skin, just because they are called "acids". Some fatty acids are exquisite for skin, nails and hair because they can penetrate and don't remain trapped on the surface. They are the just the right size and shape, so they easily penetrate. But they don't go through the hair, nails and skin, instead they concentrate inside them to provide conditioning benefits.

Fatty acids are naturally occurring chemicals found in both plants and animals, however cosmetic fatty acids most often come from plants. About a dozen types of fatty acids are found in cosmetic vegetable oils. Plants absorb nutrients through their roots and turn these nutrients into many things, including fatty acids. Fatty acids have been used since ancient times to make soaps and other useful substances. Fatty acids are the real "ingredients" being added to the cosmetic whenever a vegetable oil is used. Some vegetable oils contain low levels of other substances that are also useful, such as Vitamin E. Wheatgerm oil, for example, is one of the richest natural sources of Vitamin E. Even so, the amount of Vitamin E is very small when compared to the fatty acids. Fatty acids are in the vast majority and have the greatest impact when natural oils are added to a cosmetic product. Fatty acids are chiefly responsible for most of the physical properties of the vegetable oils. For example, the temperature at which canola oil will melt, freeze or smoke when heated is determined by the four (4) fatty acids that make up 95% of this oil. Sweet almond oil will have very similar properties, because it contains a very similar fatty acid profile, as you will see if you look at Table I.

The table lists the percentage of important fatty acids for 18 commonly used vegetable oils, most of which are widely used in cosmetics. Sweet almond is more stable and resistant to becoming rancid when compared to canola oil, due to the fatty acid called linolenic acid, the first fatty acid listed on the table. Linolenic acid is not very stable and will react with oxygen in the air to become rancid smelling and darker, making it unsuitable for cosmetics in high concentrations. It is easy to look at the chart and tell which of the oils are most likely to go rancid. The answer is Flax seed oil, since it can contain between 35-60 percent linolenic acid, while sweet almond has less than a half percent. Flaxseed is so unstable, it is not used in cosmetics, even though it has a great fatty acid profile. Therefore, high amounts of linolenic acid in a vegetable oil is undesirable. Even though wheat germ oil has 9% of this fatty acid, it also has a naturally high vitamin E content. Vitamin E's chemical name is tocopherol and it helps to prevent vegetable oils from going rancid, which is why it is often added to formulations using vegetable oils.

Vegetable Oil	Linolenic acid	Linoleic acid	Oleic Acid	Lauric acid	Myristic acid	Palmitic acid	Stearic acid
Flax seed	35-60	17-24	12-34			4-7	2-5
Hemp	24-26	54-56	11-13			5-7	1-3
Grape seed	0.1	70	16			7	4
Poppy seed	5	72	11			10	2
Wheat Germ	7-9	54-58	12-15			16-17	
Sweet Almond	0.4	20-30	62-86			4-9	1-3
Canola	8-9	8-9	63			4	2
Marula	0.1-0.7	4-7	70-78			9-12	5-8
Argan	0.3	29-36	43-49		0.1	11-15	4-7
Safflower		73-79	73-79			3-6	1-4
Sunflower		44-75	14-35			3-6	1-3
Cotton seed		42	35		0.4	20	2
Olive		4-10	65-80		0.5-1	7-16	1-3
Palm		5-11	38-52		0.5-2	32-45	2-7
Palm Kernel		0.5-2	11-19	40-52	14-18	7-9	1-3
Coconut 91°			5-8	44-52	13-19	7-11	1-3
Babussu			12-18	44-46	15-20	6-9	
Avocado			55-75			9-20	4

*Bold indicates highest percentage component

Table 1: Fatty Acid Composition of Some Common Vegetable Oils

There are about twelve types of fatty acids that can be found in all the various cosmetic vegetable oils. However, only nine of these types are found in significant amounts. Of all of these, just seven types of fatty acids are considered to be the most important cosmetic ingredients. These seven are listed in the chart; they work together to determine the physical properties of the vegetable oil. Not all vegetable oils have all seven types of fatty acids, most have between four and five. Therefore, the main difference between the various cosmetic vegetable oils are the type of fatty acids they contain. Most oils contain a blend of four or five different types of fatty acids. From this table you can learn many things about these vegetable oils, because it shows the percent and type of these most important fatty acids that are found in each oil. For instance, olive oil and palm oil have very similar composition, even though they come from very different plants. Both olive and palm oil are chiefly composed of three fatty acids, oleic acid, linoleic acid and palmitic acid.

Some vegetable oils contain only three or four types of fatty acid, while others contain up to eight. Some fatty acids occur only in certain oils. Lauric acid is a great example. This highly useful fatty acid is found only in a few types of vegetable oils, namely coconut oil and babussu oil and palm kernel oil. Each are about 50% Lauric acid, and are among the richest sources. Why is lauric acid so special? Its molecule is a great size and shape for penetration, however, this is true for all of the fatty acids shown on the chart, to varying degrees. They all are long, snake-like molecules that are wonderfully designed to penetrate. As you can see by this image. This image is a representation of a lauric acid molecule.

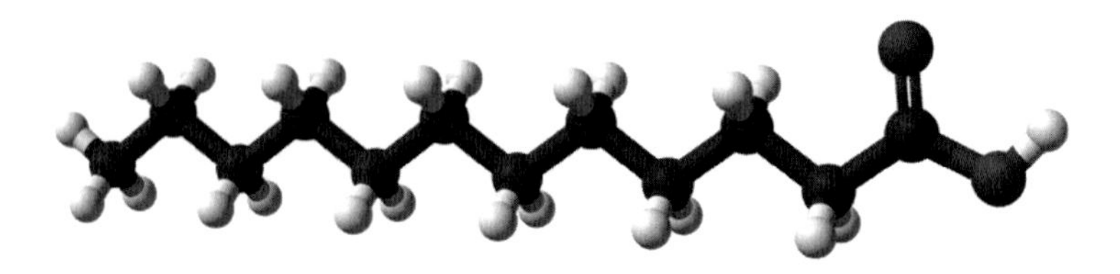

Image 17- Lauric acid molecule made of twelve carbons atoms (black), twenty-four hydrogen atoms (white) and two oxygen atoms (red).

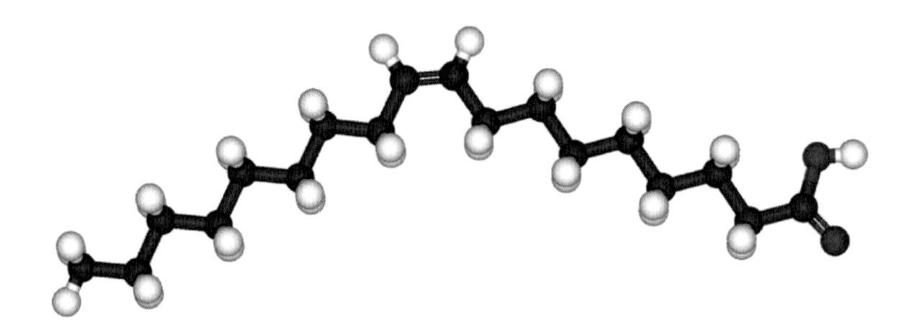

Image 18- Oleic acid molecule made of eighteen carbons atoms (black), thirty-four hydrogen atoms (white) and two oxygen atoms (red).

The backbone of a lauric acid is a chain of twelve carbon molecules, permanently linked together. They are shown as the black spheres that are connected together, that's the molecules backbone. Note the shape of a lauric acid molecule. All of the fatty acids in vegetable oils have similar snake-like structures. Along with the lauric acid, these three oils also contain 4-9% even shorter chain fatty acids with as few as eight carbons long chains, called capric acid and caprylic acid. These ingredients penetrate deeply into a shaft of hair, which is why coconut oil is often used to deeply condition hair. Most vegetable oils have longer carbon chains, some aren't straight and may have a kink or two in their chains, e.g. oleic acid. This doesn't stop them from slowly penetrating and can still probably do so even with chains up to twenty carbons long, mostly because of the molecule's shape. Of course, those with twenty carbons don't penetrate as quickly as those with eight-twelve carbons long chains. And kinks in the chain slow the penetration a little. For example, oleic acid is a bit longer and has a permanent kink right in the middle of the chain. So, it can't penetrate as quickly as lauric acid. Usually, other types of molecules this big have side branches and a bulky shape, so they usually cannot penetrate skin and just remain on the surface. The fatty acids listed in the table are between twelve and eighteen carbons long and are considered to have "medium chain lengths" and are the most useful in cosmetics.

Which fatty acids are best for nails, hair and skin? That's not so easy to answer. Lauric acid is a better penetrator, but that doesn't mean products with coconut oil are the best. This will depend on how well the oil containing product was formulated, as well as, the type and quality of the other ingredients. The same is true for other vegetable oils. Therefore, just because a product contains argan oil, doesn't mean it is a good or useful product. Product quality depends on how well the product is designed, not what cool sounding ingredients it contains. For instance, argan oil only works as a hair shine product when it is mixed with large amounts of cyclomethicone and/or dimethicone or other similar ingredients. These other substances sit on the surface of the hair to create shine, while the vegetable oils absorb to condition the hair.

The best nail oils are synergistic blends of several oils, which provide a spectrum of medium chain length fatty acids, as well as other ingredients. Avocado oil is a popular nail oil ingredient that contains high amounts of oleic acid and palmitic acid, both have great conditioning benefits. Avocado is usually combined with other vegetable oils with penetrating abilities.

But if you are curious about any vegetable oil, type the oils name into your computer's search engine, followed by the phrase "fatty acid composition" and you'll find your answer. Any good vegetable oil that contains a large amount of these fatty acids, will likely be highly beneficial when used in a well formulated product. These oils are important, but by themselves, they aren't all that! So don't be fooled. If you try to rub argan oil or any other vegetable oil into your hair, you won't like the results. Your hair would be an oily, goopy mess that would need several washes to get clean. A vegetable oil is just one of ingredients in the formula and all the ingredients work together to create a great product. You can't just buy some oil from the market and expect to get anywhere near the same benefits that will be derived from a well-designed cosmetic product.

Tips for Using Safety Data Sheets

64:2 How can you tell if a Safety Data Sheet is properly filled out and what red flag should I watch for?

Safety Data Sheets, used to be called Material Safety Data Sheets or MSDS; now the letters SDS are used. These are very important documents. Every salon should have an SDS for each of their professional products. Sometimes one SDS may cover many closely related products, e.g. nail polish of various colors or colored powders, but usually each product will have its own individual SDS. In many countries, it is a requirement to have up-to-date SDS for all professional products. For example, in the US, manufacturers are required to provide the SDSs to distributors, who are in turn required to provide them to nail professionals who in turn, are required to read and understand the information as it relates to their work and professional services. Whomever sells a professional product to a nail technician is required by US federal regulations to provide Safety Data Sheets, even if the nail professional doesn't request them. Safety Data Sheets provide important information about working safely with salon product, which is in everyone's best interest. To do so, the SDS must contain specific information and that information must be accurate. I won't list the names of each section or list all of the information that must be found on the SDS. Instead, I'll let you know how to easily find this information on the Internet. My intention is to focus on several important things to understand whenever looking at a SDS. Here are some things to check:

"Product Identification"

Make sure you have the correct SDS, the one that matches the product name exactly. Most established manufacturers are happy to provide SDS to nail technicians. However, I have seen quite a few instances where an irresponsible company did not want to provide an SDS for a product when it was requested. In one instance, a nail professional was trying to get an SDS for a nail hardener product. Rather than refuse to provide an SDS, this Florida-based nail company repeatedly sent the wrong SDS. They'd send one written for a different type of product and then claimed to have "accidentally" sent the wrong SDS. However, they repeated their mistake again when they sent replacement SDS. I suspect they were hoping the nail professional wouldn't notice, but they were wrong again.

I recently saw an SDS sent to someone from a company in China that was supposed to be for a so-called "chrome powder", but the SDS was actually for a pearl powder-which is quite a difference. BTW: none of these powders contain any "chrome", so their name is deceptive. They are actually based on pigments such as aluminum, which is an accepted cosmetic colorant. For some reason, this particular company was afraid to let people know what their ingredients were, so they provided a decoy SDS. Clearly, without the correct SDS, you can't make informed choices. Responsible companies don't play irresponsible games. In my view, if a company doesn't provide correct and accurate SDS in a timely fashion, I would not do business with them.

"Potentially Risky Ingredients and CAS#"

Contrary to what many believe, SDSs do not list all ingredients in the product. If that is what you are looking for, you will more likely find this on the product's label. The SDS is required to list only those ingredients that pose a potential risk to users. The CAS Number is the chemical identification for that ingredient, much like a fingerprint identifies a person. Look for it on the SDSs, since a CAS number is required for each potentially risky ingredient. Into your search engine type "CAS # " followed by the number found on the SDS and the ingredient name will be provided. You'll quickly discover that many chemicals have very similar names, yet they may be from entirely different chemical families and be completely unrelated to each other. Also, most ingredients have several different names. This can fool inexperienced users, which is why the CAS# is required of the SDS. Chemical names can be tricky, but CAS# provide positive identification so that you know you are researching the correct ingredient.

What about Trade Secrets?

There are many proprietary or secret formulations sold in the professional beauty industry. Companies selling these products consider their formulations to be trade secrets. You may occasionally see a trade secret claim on the SDS. I agree that there should be such a category, because this encourages innovation and helps protects these companies from copycat companies that just want to steal their ideas. However, there are specific requirements for claiming trade secret status. If the formulation qualifies for trade secret status, this allows a company to hide ONLY the identity of specific ingredients, including the name, CAS# or any other identifiers. But the SDS must still provide ALL other required information about those ingredients. All a trade secret is supposed to do is protect the ingredient identity, but does not change the SDS in any other way. If a company used trade secret claims to withhold additional required information, this would be another big red flag. Claiming special “Exemptions” or “Trade Secret” status to withhold required information is a ploy that is sometimes used to provide skimpy SDS that contain very little information. This is wrong to do and not allowed, but I have seen this done. Why would a company do something like this? Most companies would not, of course, but some do to either: Hide the use of certain ingredients, usually in order to gain an unfair advantage over their competitors who play by the rules. Or out of ignorance of their responsibilities.

Both are unacceptable excuses. The reason is often so they can fool distributors and nail technicians or perhaps so they can illegally import products that normally would be confiscated or rejected at the border because they contain unsafe or prohibited ingredients.

“Risk or Hazard Identification”

If accidental over exposure occurs, is there a risk related to skin or eye contact or from excessive inhalation? If the answer this question is yes, then the SDS will provide more detailed information about the type of risks involved.

“Signs and Symptoms Related to Overexposure” are also provided so that you spot potential problems or issues as they develop.

For instance, can the product cause skin irritation or allergy? Or can it burn the skin if not washed off completely- like callus treatments are known to do? Does inhaling excessive amounts cause sore throats, coughing or headaches? All important things to know, if you’re a working professional. If you start experiencing adverse reactions of any type, the SDS can help you and your doctor identify any potential culprits.

"Emergency First Aid"

If overexposure occurs, the SDS can recommend the proper responses. Should you wash with soap and water? Should you rinse out the eyes with warm water for five minutes or remove contaminated clothing or should you go right to the emergency room? The SDS provides these types of recommendations on a product-by-product basis. You will also find Emergency Contact Phone Numbers that you may use in case of an accident or if additional emergency medical information is needed.

"Proper Storage Conditions and Flammability"

SDS can tell you the best way to store products. This can help to maximize their shelf-life and prevent hazardous storage conditions that could lead to spills, fires, etc. If the product is highly flammable, proper storage and handling can be very important to understand to help ensure safety.

"Safe Handling"

Some of the most valuable information on the SDS explains which types of personal protective equipment (PPE) can be used to avoid overexposure and help ensure safe handling. Wearing specific types of gloves and encouraging the use of protective safety eye wear are common recommendations. Ventilation recommendations are also found on the SDS and this information should be carefully heeded. For many salon products, the use of proper ventilation is an important requirement. The SDS can help identify which products are inhalation risks and how to minimize exposure to keep them at safe levels. Many salons do not pay proper attention to these important issues, but the information provided on the SDS can help ensure that nail technicians have long, safe and successful careers. Working safely is much easier to do, if you have the correct information.

Of course, there is MUCH more information on the SDS, more than ever before. Much of it is not useful for nail technicians. Why is that? SDS are made for everyone, including firefighters, truckers, medical doctors, manufacturing workers and scientists. A typical SDS will contain sixteen different sections, each focused on one important aspect of safety, for example, what to know when cleaning up accidental spills. But don't be intimidated by all this extra information, instead seek out the useful information I described above. As the information on the SDS expands, so does their usefulness, but now each product's sheet is typically between four to fifteen pages long for each SDS. Here's another red flag, if you are sent a one-page SDS, you should be suspicious that the SDS is not complete. In some unusual cases, it may be possible for a properly filled out SDS to be only 2 pages, this it is becoming increasingly unlikely. This is due to all the information the SDS must contain.

Other than the information I've already discussed, they must contain information about the physical and chemical properties, proper disposal, shipping and transportation, ecological impact and a list of applicable regulations. That's why in my opinion, a short SDS is likely an incomplete SDS. It is the responsibility of every cosmetic manufacturer, large or small, to fully and faithfully comply with these regulations by providing SDSs that are complete and contain all the required information. It is the responsibility of all cosmetic manufacturers and their distributors to ensure the most up-to-date SDS are consistently provided to the end user, which is the nail technician. And these SDS are supposed to be provided even if the nail technician did not ask for them, but that often doesn't happen because of the sheer volume of paper that would be consumed. Therefore, it is wise to ask for them. Many companies now offer their SDSs on-line at their website.

It is also the salon owner's responsibility to have SDS for all professional products on-site and available for inspection during normal business hours and to provide training on how to read and understand these important tools. They don't have to be printed copies, just so a digital copy can be produced on demand when anyone asks for them during normal business hours or in case of emergency. Finally, it is the responsibility of all nail technicians to understand how to work safely and wisely with their professional products. A properly filled out SDS is one of the most powerful tools a nail technician has to help ensure they are working safely. If you want to learn more about Safety Data Sheets. Type the following into a computer search engine, "Understanding SDS Sheets", and you'll be on your way to learning more and working safely.

74:5 I work in my own home based salon in a small town. My problem is that clients share a lot! I focus on my work and can't always remember the details they share, but some clients feel disrespected when I don't remember something personal. I can't keep up all the time with everyone's lives, but they want to chill and be heard. What's a nice way to say this?

I really love your question, since it is a common salon problem. Even so, it is a tough question because many clients love to use their nail technician as a sounding board and they think that is part of the service. But it brings to mind something interesting a nail technician in the UK said to me recently. She told me that since they read my latest book, her clients don't want to talk about themselves as much, but instead, want to hear more about their nails. They love when she busts myths and it starts a conversation about something other than their kids or husband. Personally, I think it makes the nail technicians more valuable and the clients more loyal, since they know their nail technician is more knowledgeable than the most. So, there are many unexpected benefits from being more knowledgeable about nails and products... so keep learning and sharing.

Additional Special Topics

Topic 1 - Does nail polish transmit pathogens to cause infections?

Two independent scientific studies confirm that nail polish products do not harbor microbes, so it is not likely that any pathogen could be spread to cause infections. Water-based cosmetic products typically contain preservatives to prevent microbial contamination, when they are sold in multi-use packages. Organic, solvent based nail polish products (e.g. nail polish, lacquers, enamels, varnish, base coats, top coats) provide a hostile environment for pathogens and prevents microbial contamination even when the product and brush are used repeatedly (double dipped) and therefore additional preservatives are not needed or added. Professional use nail polish products are essentially water-free, organic, solvent based mixtures of colorants, film formers and other additives. Some have questioned the use of nail polish products in salons over concerns of a possible increased risk for transmitting microbes when a polish brush is reused on multiple clients.

Several independent scientific studies, conducted by the Nail Manufacturer's Council on Safety (NMC) of the Professional Beauty Association, demonstrate that organic solvent based nail polish products aggressively kill any microbes that may be inadvertently picked up by a nail polish brush, therefore repeated use of nail polish products does NOT pose an infection risk for salon clients.

In the first study, unopened nail polish products were intentionally mixed with high concentrations of seven common microorganisms associated with nail or skin infections, followed by laboratory testing to determine if any of these microbes could survive or reproduce. The products tested contain the volatile organic solvents used almost universally by all manufacturers (e.g. ethyl acetate, butyl acetate, isopropyl alcohol, etc.) which are typically make up 60-70% of nail polish formulation. These nail polish products were identically inoculated with a high concentration of live microorganisms, then tested immediately and periodically for fourteen (14) days to ensure there was no regrowth of microorganisms.

Results: In all the nail polish products tested, the introduced microbes were rapidly destroyed and there was no regrowth, even after fourteen (14) days.

In a second study, twenty (20) half-used nail polish bottles, representing eight commonly used salon nail polish brands, were repurchased from ten (10) nail salons. The collected nail polish containers were less than half-full, indicating they were used on approximately twenty-five (25) clients before repurchase. The collected containers were then submitted to an independent laboratory for testing.

Conclusions: Professional-use nail polish products which do not contain water as an ingredient and are mostly made of organic solvents can rapidly destroy microbes; which explains why these products don't require traditional cosmetic preservatives. These results demonstrate that microbes cannot live in water-free nail polish products and any microbes accidentally introduced would be rapidly destroyed. These results also demonstrate that a properly used, professional-use nail polish product may be safely applied to multiple salon clients because the organic solvents are an effective deterrent against microbial contamination.

If you want more details about these studies you can find them in the Nail Manufacturer's Council on Safety's education brochure,

"Investigation of the Potential for Microbial Contamination in Nail Polish" https://probeauty.org/docs/nmc/Micro_Contamination_of_Polish.pdf

Topic 2 - The Game of Fear: Cosmetic Activism Gone Wild!

I received yet another e-mail from an anti-cosmetic activist group asking me to donate money to their "cause". This is the third such e-mail in the last few weeks. It does not matter which group sends the e-mail, they all start the same way; telling us how lucky we are that "they" are working to save us from all the supposedly dangerous "chemicals" in cosmetics. You know, the ones that are supposedly killing babies and everybody else too. Of course, this is just pure nonsense, but that doesn't seem to matter to these fear-based groups. They intentionally misquote studies and distort statistics to support the bogus information they spread on a regular basis.

According to these groups, everybody who uses cosmetics or personal care products is at risk of great harm! What? Cosmetics? Apparently, it doesn't matter that these types of products have a very long history of safe use and overall impressive safety record. The facts are, cosmetics and personal care products are among the safest products that consumers can purchase according to the U.S. FDA and Health Canada. These fear-mongering groups do NOT want you to know the truth. Instead, they use tricky and deceptive claims like "*may cause...*" or "*it is believed that...*" or "*could be linked to...*" They resort to these misleading phrases because they lack real proof to justify their claims, so instead rely on their bag of tricks.

What else can they do? Without any solid evidence, they can't debate the facts with knowledgeable scientists. Instead, they use the Internet and the news media to launch unfounded attacks. They pretend to be champions of truth, when instead they are the exact opposite. Hair shampoo and conditioners do not cause cancer or birth defects. Lipsticks are extremely safe and do not contain harmful amounts of lead. Nail polish has been safely used for 80 years, until these groups created a hoax to make people believe it is suddenly unsafe. The list goes on and on. They've manufactured the false notion that hidden dangers lurk in these types of products and their scare tactics have made a lot of money for them! Look at their e-mails and you'll quickly discover the “Donate Here” link. Now they are using mobile apps to collect even more money, so it's no surprise they rake in huge amounts of money!

My advice is to never donate to any group that focuses on spreading fear and misinformation about cosmetics and their ingredients. They will just continue needlessly frightening your friends and clients. Please don't support such groups or visit their websites. Fear-based advocacy groups are self-serving and don't have the public's best interest at heart. Most are money making machines that seek to control the political process so they can promote their own irresponsible agendas. Don't be fooled. They don't really care about any of us. Much of this money will be used to promote misguided regulations that aren't needed and if passed would cause more harm than good. So please don't donate to any “fear-based” activist groups. Instead, tell them you don't appreciate their exaggeration and distortions and ask your friends/clients to say the same. These groups are unlikely to stop their trickery, as long as it is lucrative to frighten the public.

Topic 3 -
Do you know of a chemical-free product? Then you can make $1.1 million (£1million) and here’s how.

Want to make £1million (1.1 million US Dollars) the easy way? All you must do is identify a truly chemical-free product. That should be easy with so many companies claiming to have “chemical-free” products. A £1million bounty was offered for UK's first chemical-free product. The prize has been available since 2010, but so far no one has claimed the money? Hmmm, I wonder why? Could it be that those who make such claims know they are tricking the public for personal gain and profit? To expose this trickery, the Royal Society of Chemistry (RSC) announced the award a £1 million bounty to the first person who can crack this impossible task and create a product that is truly 100% chemical-free. Why did this prestigious society offer this challenge? Research by the UK's cosmetic and toiletries industry reveals 52% of women and 37% of men actively seek out chemical-free products, demonstrating a deep-seated public confusion about chemicals and how we interact with them in daily life.

The facts are, the clear majority of chemicals that people come into contact with each day are not only safe, they are beneficial and essential for life. Our bodies are constantly manufacturing chemicals. For example, proteins are chemicals made by other chemicals called amino acids. This process requires use of the powerful chemical solvent known as water. Yes, water and air are chemicals, as well. Anything that contains any elements are automatically a chemical.

Read more about the challenge, as well as the deception and trickery used by some to fool the public into falsely believing that chemicals are dangerous by clicking the link www.rsc.org/AboutUs/News/PressReleases/2010/CTPA100ChemicalFree.asp

Topic 4 - Keep Counterfeit Products Out of Salons

In my view, two of the biggest problems facing the professional beauty industry could best be solved by salon professionals. These two big problems are tightly interlinked; you might say they're "birds of a feather". It is widely known that diverting professional products into salons through unauthorized channels of distribution hurts the salon industry and can take money out of everyone's pocket, except for those of the diverters. Unauthorized distribution of professional products is referred to as the "grey market". This practice is technically legal, there's little a product manufacturer can do to stop someone from selling professional salon products in supermarkets, swap meets, on the internet, etc. Even so, the original product manufacturer is often unfairly blamed by salon professionals for "going retail".

The facts are, grey market distributors also are likely sources of counterfeit products, which is the other big problem the industry faces. Some may not see much harm in buying a counterfeit Gucci purse or fake Rolex. At most your purse falls apart after a few months or several fake "diamonds" suddenly pop off your watch. This is quite a different experience from what can happen when professional salon products are counterfeited. These may have the potential to injure or harm salon professionals or their clients. Remember that counterfeit products are designed to "mimic", not "duplicate". Counterfeit professional salon products are not likely to be manufactured using proper quality control, nor will they undergo the same quality and safety testing performed by the original product manufacturer. Counterfeit products are likely to contain substitute ingredients that are less costly than ingredients used by the original manufacturer and may not be safe replacement choices. Counterfeit products have been known to cause serious adverse skin reactions and burns, just to name a few of the potential problems associated with their use.

While visiting a trade show in Ireland, a nail professional approached me with a very interesting story. She found her favorite UV gel manicure product on the Internet at a lower cost than what her authorized distributor was charging and thought she was getting a great deal. When the product arrived, she noticed the packaging looked slightly different. A printed piece of paper inside announced, "*We've has recently change this packaging*", which insinuated there was no need to be concerned, but clearly the grammar errors should be a red flag indicating that something is wrong. During product application, she noticed the odor was different and the product didn't seem to cure very well. Clearly, she had purchased a counterfeit version of this popular brand. It's no surprise that popular brands are often the ones that are counterfeited; so beware if you buy your favorite products from an unauthorized distributor or source, e.g. the Internet. The most obvious red flag should be the price, especially when the price is surprisingly low. Besides misspelling on labels and improper grammar, poor-quality logos or art work is another red flag. If you're unsure about a distributor, contact the original product manufacturer to determine if they are an authorized source.

If you suspect that a professional salon product you purchase is counterfeit, I recommend reporting the seller to the local authorities, as well as informing the original product manufacturer. This is the best way for salon professionals to put a stop to this unfair and risky practice. Salons can play an important role in stopping these two big problems simply by ensuring all professional salon products are purchased from an authorized distributor or directly from the original manufacturer. Counterfeit professional salon products don't just harm the industry; these fake products can harm people. Help keep them out of salons.

Topic 5 - Reducing Inhalation Exposure to Nail Salon Products

If you know how, it is easy to improve the quality of your salon's air and minimize inhalation exposure to potentially harmful substances. Yes, there are things you can do to avoid excessive inhalation of dusts or vapors. By taking the right steps you can improve the salon environment for customers and create a safer, more pleasant workplace for salon professionals by removing vapors or dusts from breathing air.

Reducing exposure to vapors and dusts can also help sensitive individuals avoid symptoms such as irritated eyes, nose or throat, headaches, difficulty breathing, nervousness or drowsiness. Each of these can be related to poor ventilation or ventilation that's not appropriate for the services being performed in the salon. What? Even drowsiness? Sure. We normally exhale carbon dioxide (CO_2) and if the ventilation is poor, levels can build up in the salon and make you feel tired, lethargic and can even lower your performance and decision-making skills. Poor ventilation increases the air concentrations of product vapors and dusts as well, which is also important to avoid. For these reasons, I recommend that nail professionals and salons take the appropriate steps to ensure that they have proper and appropriate ventilation in the salon.

Just about every substance on Earth has both a safe and potentially unsafe level of exposure. Salon vapors and dusts are no exception. In general, the vapor levels in a properly ventilated salon are well within safe limits. However, not all salons have proper ventilation and those that don't, often do not understand its importance or they may not understand the correct steps to take or even where to begin. There is a lot of misinformation about salon ventilation.

For instance, it's a myth that you can determine the safety of a nail product by how it smells. Some salon professionals mistakenly believe that ventilation systems are solely for controlling strong odors, when in fact, odors are not the reason for ventilating. The odor of a substance does not indicate whether it is safe or harmful. Dirty socks and baby diapers provide examples. Although these may not smell very good, they certainly aren't harmful to breath nor are their odors dangerous. Some vapors have very little odor, yet they should also be controlled and kept at safe levels. Fragrances smell wonderful, yet some people are sensitive to inhaling excessive amounts. So, it is important to have a good understanding of these issues, as well as, to understand which steps to take to keep vapors and dusts under control and within safe levels.

One of my many responsibilities as Co-Chair of the Nail Manufacturer's Council on Safety (NMC) is to work with other nail industry scientists and experts to create informational brochures that help nail professionals understand how to work more safely with professional nail products. Recently, I worked with a team of NMC and ventilation industry experts to create a significant new update to an existing NMC brochure. This update provides some new and important information. It's entitled *Guidelines for Controlling and Minimizing Inhalation Exposure to Nail Products*". You can find a copy of this brochure at this link, http://www.schoonscientific.com/wp-content/uploads/2016/08/Guidelines-on-Inhalation-Exposure_ENG.pdf.

In it you will discover many useful ideas for improving salon air quality. The brochure includes general explanations about the types of ventilation systems available, as well as their usefulness in the salon, how to select and properly use dust masks and a list of other important tips and suggestions for minimizing and lowering exposures to safe levels. I encourage you to take the time to download and read this useful guide. It's an important way to help safeguard your heath and to provide yourself with the peace of mind that comes with having the right information needed to make wise choices that improve the quality of your breathing air.

Topic 6 - The FDA Believes Fluorescent and LED-style UV Nail Lamps are Safe.

At last the FDA (aka US Food and Drug Administration), has spoken out and VERIFIED that UV nail lamps are surprisingly safe (see link below). Hopefully, this will address misguided concerns from the likes of Dr. Oz and other naysayers. Hopefully it will silence the Fake News around this issue. Keep a copy handy to share when needed. (Good news travels slowly and fear-mongers don't give up easily).

Here is a quote from this article. Please share this information with all your nail friends and clients and on other social media sites.

“...the FDA views nail curing lamps as low risk when used as directed by the label. For example, a 2013 published study indicated that—even for the worst-case lamp that was evaluated—36 minutes of daily exposure to this lamp was below the occupational exposure limits for UV radiation” and “To date, the FDA has not received any reports of burns or skin cancer attributed to these lamps.”

In other words, even putting a hand in a UV nail lamp every single day for 10 minutes, it’s still safe. That’s because UV nail lamps are designed to be safe and have ALWAYS been safe- despite what many fake experts have wrongfully claimed. Lots of other good info in this article as well, so please read it.

Also, if you’re one of the misinformed who don’t know that LED nail lamps emit MORE UV energy, don’t be shocked, but it’s true. That’s how they cure faster. Even so, the shorter exposure times mean the skin is exposed to about the same very low levels of UV that are created by traditional UV fluorescent-style nail lamps. LED-style nail lamps were also a part of the study cited by the FDA and are ALSO considered to be just as safe as traditional UV nail lamps. So, please don't be fooled into thinking they are safer.

BTW: The article talks about UV radiation which is NOT the same as “radioactive radiation”. Heat, light and UV energy all “radiate” from their source of emission. This word simply means to “move away” from the source. Automobile radiators are not radioactive either. Don’t be confused by this word. UV radiation is the same thing as UV energy, aka UV light. Even visible light radiates from a bulb and is called visible radiation.

In short, UV nail lamps, including fluorescent and LED-style, are safe as used! Here is a link to this article:

http://www.schoonscientific.com/resource/
latest-fda-uv-nail-lamps-safety/

Topic 7 -
More on the the risks of using e-files on the skin around the nail plate.

In Volume II, I wrote about my concerns for using e-files on the skin surrounding the nail plate, (aka the Russian or machine manicure) and why it should not be performed, when this includes using an e-file to abrade the skin. This opinion was based on over two decades of experience studying nail salons; products, services and common practices. Please read my comments in Volume II, but I wanted to add the following information for your consideration. In my view, e-files are very risky to use on the living skin. The study and images below support of my conclusions. You can see from the images shown below that microscopic damage occurs to the nail plate as well as the living tissue surrounding the plate. Whenever the skin is broken like this, infections become a very real possibility.

This information was supplied by the Russian nail scientist and physician, Vitaly Solomonoff:

“The background of the so called “Russian manicure“ is an attempt of e-file distributors to sell their units in Russia. That is why it has been widely advertised as a safe and “healthy” alternative to clipping the nail “cuticles”. We, at my company have always stood against this practice unless NT has at least basic medical education and full understanding of Anatomy and Physiology of the nail unit. We have seriously studied more than 300 cases. The results are as follow – 91% of clients who constantly get the Russian manicure have symptoms of the damaged nail matrix or nail bed. It is important to note- those symptoms may not appear right after manicure is performed, the first symptoms occur months after due to the constant repetitive traumatization of the cuticle/matrix area. Symptoms include signs of matrix/nail dystrophy from splitting, horizontal ridges, slow nail growth, to the painful neuropathy and high sensitivity. We have also discovered that infectious inflammations are a common issue in those clients who have compromised immune system, diabetics, etc. This type of inflammation occurs even when the manicure performed with ideally sterilized implements. We strongly believe that micro cracks of the skin are inseminated with bacteria during few hours after the procedure.

Only 9% of the cases have been determined as safe and "successful" during this 38-month study and all the successful client experiences from NTs with basic medical education. We concluded from these results that that deep understanding of the processes concerning living skin helps to ensure a correct technique for this type of manicure. I agree with Doug that this kind of procedure is a type of microdermabrasion and requires the special knowledge, training and full understanding of what happens to the skin and nails under attack of the vibrating, sharp bit of e-files. Imperceptible vibration always occurs, even with high end e-files and affects the highly sensitive nail matrix area, which leads to the delay in seeing problems. The technique that may look safe and easy can bring troubles in the future." See related images below.

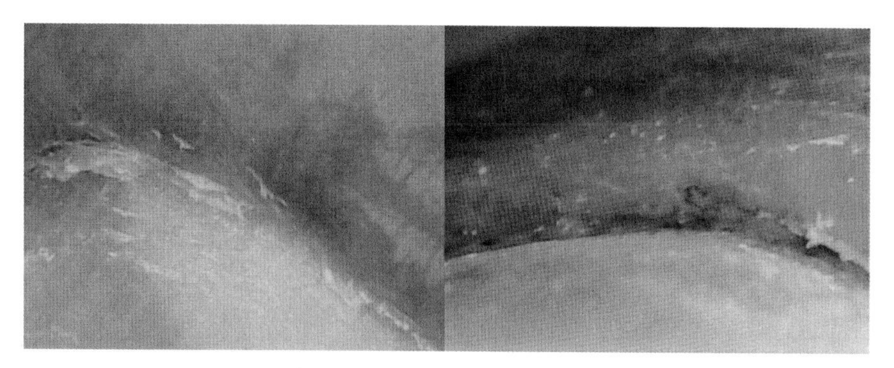

Image 19a - Nail plate layers damaged with an e-file bit in the cuticle area

Image 19b - Free edge point of view. Microscope shows multiple micro-wounds which are invisible to the naked eye.

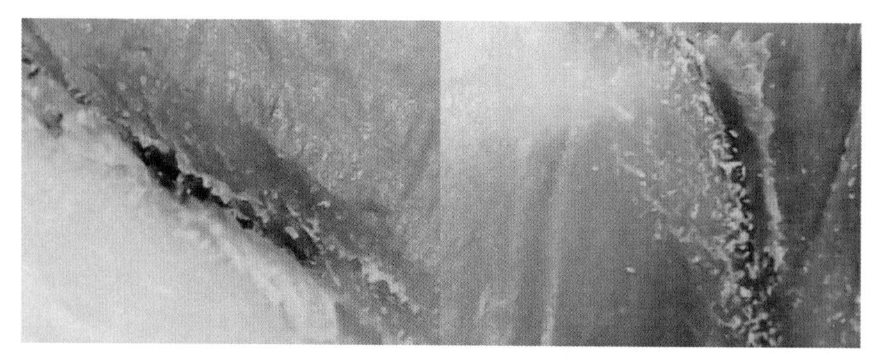

Image 19c - Tiny micro-wound/laceration of the nail fold' invisible to the naked eye.

Image 19d - Lacerations and damage of the living skin and visible damage of the nail plate.

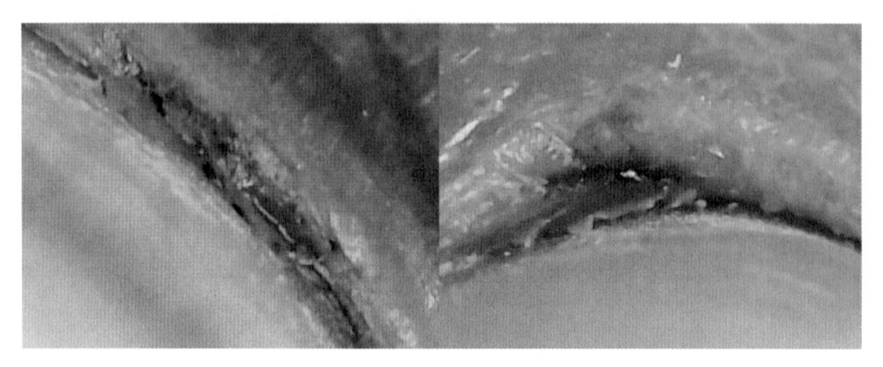

Image 19e- Proximal nail fold with microscopic damage caused by e-files.

Image 19f- More microscopic damage of the proximal nail fold.

Image 19 a-f are a series of research photos that demonstrate examples of typical damage caused by electric files when used to remove the cuticle from the nail plate or hardened tissue on the nail folds that surround the nail plate.

(Photo credit: Vitaly Solomonov Dermatologist/Cosmetic Chemist, Cleanestiq Labs LLC. Philadelphia, PA)

For more information about this risky technique and why it should be avoided, see Volume II, page 41.

Topic 8 - Controlling Dusts Created by E-Files

Problems may occur when nail technicians ignore excessive inhalation of dusts (or vapors). Filing can create a lot of dusts that ends up in the breathing zone of the salon worker. Disposable dust masks can be used to minimize inhalation of dusts, as can source capture ventilation. Dust masks should never be used in place of proper ventilation. Even so, when properly used, the correct mask can be an important way to prevent inhalation of excessive amounts of dust particles. They are especially useful to those with pre-existing asthma, allergies, or other breathing related conditions. The best dust masks are those specifically designed for blocking dusts. These are thicker than most masks and better able to block most of the fine particles that attempt to penetrate the mask. For any dust mask to work well it must fit well. Better quality dust masks will fit and seal more securely and comfortably and do a better job overall. Dispose of these masks regularly and follow the manufacturer's directions for use. Avoid using surgical/doctor-type masks since these will not provide enough protection and should not be used in the salon setting. Doctor-type masks may help prevent the spread of germs, but are not suitable for salon work.

I also recommend using electric file nail oil when filing to reduce the amounts of dust in the air. These nail oils are especially designed to work with e-files and this is a highly effective way to significantly reduce the amount of fine dusts in the salon air. The image shows how easily dust spreads to the breathing zone when electric file nail oils are not used. The chart indicates the typical particle size of nail dusts, when hand and e-filing are compared. Note that e-files create much more of the smaller size dust particles as shown by the range of the hand file and e-file bars on the chart. The larger particles created by hand filling, tend to clump together and quickly fall out of the air to cover the tabletop, floor, as well as to collect in the hair or inside the collar. They make a mess, but at least they aren't in the air, so they can't be breathed. Note that e-files create much more of the smallest dust particles, which are much lighter and can remain suspended in the salon air for hours. These extra-small dusts created by e-filing are a greater inhalation risk, so they are more important to control and avoid. These smaller particles are easier to inhale and get pulled deeper into the lungs. Of course, the lungs are good at handling dusts and getting rid of them. They have to be, we live in dusty world and this chart shows that many even smaller practices are often inhaled, but that's not a reason to ignore this issue.

Whatever method you use to file, it is important to keep your work area clean and to remove all visible dusts, but be aware of invisible dusts. Those smaller than 10 microns will not be easily seen and will float around the salon so that everyone can breathe them. Luckily, these invisible dusts are easy to control by, 1). Good housekeeping practices to control dusts, 2). Use of appropriate ventilation, e.g. "source capture" systems that remove dust from the source, as they form, 3). Use of an electric file nail oil. 4). Proper use of a suitable dust mask.

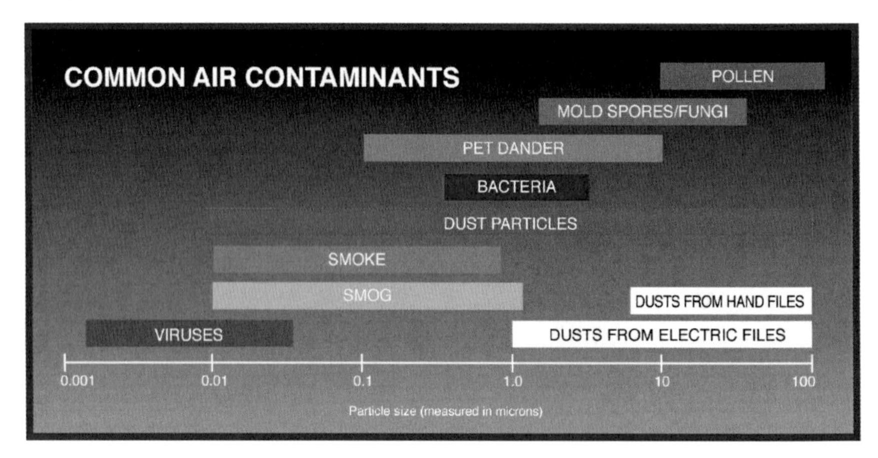

Image 20: Particle size of several air contaminants in relation to dusts created by hand filing and electric files.

Image 21a: A large volume of very small dusts particles are created by electric files.

*Image 21b: A 3D image of the dusts created by e-files, **Red/Cyan** 3D glasses required.*

Index

Appendix Links

Free 3D "Anaglyph" Glasses Offer for Readers:

Rainbow Symphony, a specialty eye glass maker, has very kindly offered to send to all readers of this book one (1) free pair of 3D glasses which will allow viewing of the 3D images.

To get your free 3D glasses you must "exactly" follow the directions below:

Send a "Self-Addressed Stamped Envelope (SASE) with the proper return postage on the envelope to: Rainbow Symphony, Inc., 6860 Canby Ave. Suite 120, Reseda, CA, USA 91335. Mail the SASE inside the letter that you send to Rainbow Symphony. They will use your SASE to return your free glasses.

Make sure to specify the type of glasses needed as, "Paper Anaglyph **Red**/Cyan glasses". This is important to include.

If you do not include a "Self-addressed Stamped Envelope (SASE), with enough postage they won't have any way to send you the glasses, so this is very important. As of the printing of this book, US postage stamps are 0.49 cents.

Check your local post office for postage rates to your location if you are unsure. The glasses weigh no more than a standard letter with two sheets of paper included, so they are very light-weight.

This offer is good for anyone, anywhere in the world, but those living outside the United States MUST ensure the SASE has the proper amount of postage or the glasses cannot be sent to you.

For more information, see https://www.rainbowsymphony.com/

Note: **Red/Cyan** Anaglyph glasses may also be purchased on-line everywhere and found in many novelty shops, but the Left Lense must be “Red” and the Right Lense must be “Cyan” color.

Acknowledgements

Technical Editor:

Holly L Schippers
FingerNailFixer®
http://www.nailsmag.com/bloglist/fingernailfixer
http://youtube.com/fingernailfixer
http://facebook.com/fingernailfixer
http://instagram.com/fingernailfixer

Photographer:

Paul Rollins
http://paulrollinsphotography.com/

Back Cover Photograph:

Judy Landis-Storm

Made in the USA
Middletown, DE
22 September 2024

61291478R00097